Marine Mammals

Edited by Hussein Abdelhay Essayed Kaoud

Published in London, United Kingdom

Marine Mammals
http://dx.doi.org/10.5772/intechopen.98133
Edited by Hussein Abdelhay Essayed Kaoud

Contributors
Adib Ali Saad, Ilene Mahfoud, Elena Vecino, Noelia Ruzafa, Xandra Pereiro, Sansar C. Sharma, Reyna Cristina Collí-Dulá, Ixchel Mariel Ruiz-Hernández, Md. Muzammel Hossain, Tetsuya Endo, Mari Kobayashi, Luis A. Pastene, Mutsuo Goto, Mioko Taguchi, Yoshihiro Fujise, Hussein Abdelhay Essayed Kaoud

First published in London, United Kingdom, 2022 by IntechOpen
IntechOpen is the global imprint of INTECHOPEN LIMITED, registered in England and Wales, registration number: 11086078, 5 Princes Gate Court, London, SW7 2QJ, United Kingdom

British Library Cataloguing-in-Publication Data
A catalogue record for this book is available from the British Library

Additional hard and PDF copies can be obtained from orders@intechopen.com

Marine Mammals
Edited by Hussein Abdelhay Essayed Kaoud
p. cm.
Print ISBN 978-1-80355-489-1
Online ISBN 978-1-80355-490-7
eBook (PDF) ISBN 978-1-80355-491-4

Meet the editor

Dr. Hussein Kaoud has a Ph.D. and DSc and has completed veterinary fellowships. He is currently a Full Professor of Preventive Medicine, at Cairo University, Egypt, where he was previously Chairman of the Department of Preventive Medicine. He has given lectures in Molecular Epidemiology and Biotechnology at different universities and has been a reviewer and editor for several scientific journals. His research interests include molecular biology and advanced technology in basic life sciences. He has more than 300 publications to his credit, including several international books. He has received one international award, twenty Cairo University International Publication Awards, and the Appreciation Award in Advanced Technological Sciences from the same university. Dr. Kaoud has supervised and examined many medical dissertations.

Contents

Preface

There are 130 described species of marine mammals living within the Earth's oceans. These include cetaceans (whales and dolphins), pinnipeds (33 species of aquatic fin-footed mammals, such as fur seals and sea lions), sirenians (sea-cows or sirenians), sea otters (the smallest marine mammals), and polar bears.

This book offers a broad perspective on marine mammal species and populations considered to be most at risk due to human activities. It emphasizes the importance of understanding the essential biology and ecology of marine mammals to assess the correlates and causes of extinction and to implement science-based conservation.

The book contains seven chapters.

Chapter 1: "Introductory Chapter: Marine Mammals of the World"

This chapter discusses different types of marine mammals and their importance to the ecosystem of the world's oceans and food cycles. Additionally, the chapter addresses the threats facing marine mammals and attempts to answer the questions: Why is it important to safeguard marine mammals? Why is marine conservation important? How can we protect marine mammals? Finally, the chapter illustrates global hotspots where marine mammal species are at risk of extinction and examines the geographic distributions of the leading human impacts.

Chapter 2: "Phylogeny and Population Genetic Structure of Minke Whales Worldwide: A Review of Recent Studies"

This chapter reviews the genetic studies on minke whales. The review is organized by topic, e.g., those studies focused on phylogeny and other matters most relevant for taxonomy, and those focused on population genetic structure.

Chapter 3: "Applications of Omics Approaches to Decipher the Impact of Contaminants in Dolphins"

This chapter explains the importance of omic technologies in the field of marine genomics with the advent of omic technologies (genomic, transcriptomic, proteomic, metabolomic, and lipidomic).

Chapter 4: "Typical Changes in Carbon and Nitrogen Stable Isotope Ratios and Mercury Concentration during the Lactation of Marine Mammals"

This chapter deals with lactation in marine mammals.

Chapter 5: "Perspective Chapter: Status of Dolphin in the Maritime Area of Bangladesh"

This chapter discusses dolphin survival rates and causes of movement reduction in Bangladesh. The dreadful conditions of coastal habitats can have major impacts on dolphin population and distribution.

Chapter 6: "Marine Mammals in Syria"

Syrian marine water is one of the least studied areas for cetaceans in the Mediterranean Sea. Lack of basic knowledge such as species composition and habitat makes it impossible to develop effective conservation measures. The survey in this chapter was carried out along the Syrian coasts.

Chapter 7: "How Do Whales See?"

This chapter studies the eyes of two whales: *Balaenoptera physalus* and *B. borealis*. It presents anatomical, histological, immunohistochemical, and ultrastructural studies of the eyes of both types of whales.

The book is characterized by precision and care and ensures that there aren't any errors – irrespective of how small – within the scales, colorations, and details of the drawings. Fortunately (or otherwise), many of these rare species are still unknown to most people – and may remain unknown forever.

I gratefully acknowledge the help provided by all authors that have contributed to the publication of this volume. I am also thankful to IntechOpen for initiating this project.

Hussein Abdelhay Essayed Kaoud
Professor,
Faculty of Veterinary Medicine,
Cairo University,
Giza, Egypt

Chapter 1

Introductory Chapter: Marine Mammals of the World

Hussein Abdelhay Essayed Kaoud

1. Introduction

1.1 Definition

Marine mammals are, in short, catalytic species. Thus, knowledge of areas that are important for them will facilitate the balancing of human uses of the ocean with the imperative of conserving marine biodiversity.

Marine mammals are highly consumed animals of marine creatures production at the different levels specially utmost trophic type from initial production as sea-cows or sirenians, are an order of fully aquatic, herbivorous mammals) to predatory fish and some other types of marine mammals, such as the polar bear, some whales, and dolphins.

2. Role of marine mammals in ecosystems

Because of their large body size and abundance, they are thought to own a serious influence on the structure and performance of some marine communities.

In marine ecosystems, marine mammals play an important role in balancing and regulating the dynamics of production in the ecosystems as the following:

1. They provide a context to regulate and to judge the potential influence of their predation on prey populations.

2. They regulate and balance the structure of marine populations.

3. They impact or influence the variation in prey populations that harvesting by humans.

4. They regulate the dynamics of marine mammals [1] during environmental changes.

3. Threats due to human activities

Human activities represent a big threat to words marine mammals. These activities comprise "Bycatch" in marine media, collision with vessels, reduction of prey resources, and climatic changes on the planet. The danger of water pollution, excessive hunting or fish harvesting, diseases, and habitat destroying or degradation and loss.

For the following reasons, marine mammals represent a complex problem and challenges for conservation:

1. Lacking data concerning marine mammal bycatch.

2. Lacking data concerning marine mammal species-specific data.

3. Lacking marine studies and research.

Bycatch could be a complex, global issue that threatens the sustainability and resiliency of our fishing communities, economies, and ocean ecosystems. Bycatch of protected species, like sea turtles and marine mammals, remains a big threat to recovering dwindling populations [2, 3].

Global data on marine mammal bycatch is mostly lacking, particularly species-specific data. For these reasons, marine mammals present an array of issues and challenges for conservation and management.

Bycatch could be a complex, global issue that threatens the sustainability and resiliency of our fishing communities, economies, and ocean ecosystems. Bycatch of protected species, like sea turtles and marine mammals, remains a big threat to recovering dwindling populations.

4. Marine mammals in the world

Marine mammals are a diverse group of species that include:

1. Cetaceans

2. Pinnipeds

3. Sirenians

4. Sea otters

5. Polar bears

4.1 Cetaceans

Cetaceans are a cosmopolitan and diverse clade of aquatic mammals; they include whales, dolphins, and porpoises. They are flesh-eating (carnivorous) and most of them are found in the oceans.

Cetaceans are differed in size and weight, the length ranged from 1 m in Maui's dolphin to 29.9 m as in blue whale, and the weight ranged from 50 kg to 173,000 kg.

4.1.1 Whales

- There are now 15 species (**Figure 1**) and around 89 extant species [3].

4.1.2 Dolphins

- Families Delphinidae (the oceanic dolphins) (**Figure 2**) [2]

Figure 1.
Types of whales in the world.

Figure 2.
Types of dolphins in the world.

- Platanistidae and Iniidae (River dolphins)
- New World:
  - Bottlenose dolphin
  - Common dolphin
  - La Plata dolphin (disambiguation) dolphin,

40 species of dolphins within the Delphinidae, 6 are commonly called whales, including the killer and also the pilot whales.

Most dolphins are small, measuring 3 meters (10 feet) long, and have spindle-shaped bodies, beaklike snouts (rostrums), and easy needlelike teeth. a number of these cetaceans are occasionally called porpoises.

4.2 Pinnipeds

The pinnipeds are a group of 33 species of aquatic fin-footed mammals composing three families (**Figure 3**) [4]:

Figure 3.
The pinnipeds are a bunch of 33 species of aquatic fin-footed mammals composing three families.

DOI: http://dx.doi.org/10.5772/intechopen.105110

- Verity seals (family Phocidae),
- The fur seals and sea lions (family Otariidae),
- The walrus (family Odobenidae).

Pinnipeds live in some inland or tropical freshwater systems and rich marine environments.

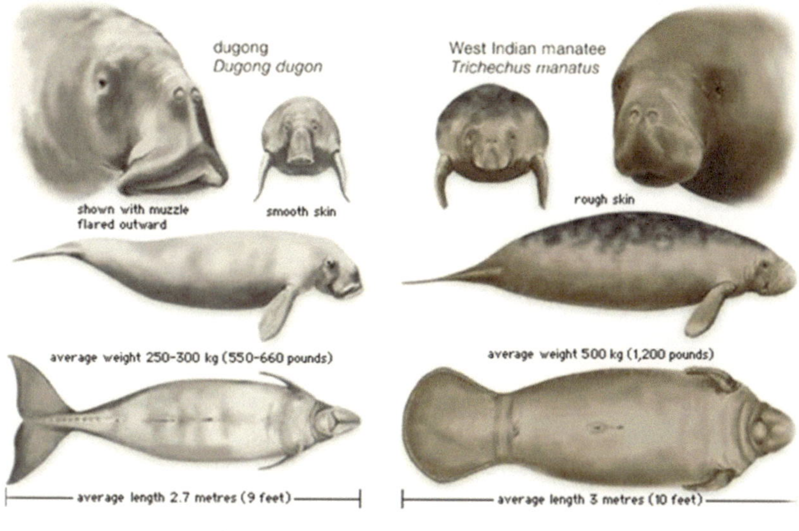

Figure 4.
The Sirenia commonly spoken as sea-cows or sirenians.

Figure 5.
The sea otter.

Figure 6.
*The polar bear (*Ursus maritimus*).*

4.3 Sirenians

The Sirenia (**Figure 4**) commonly spoken as sea-cows or sirenians are an order of fully aquatic, herbivorous mammals that inhabit swamps, rivers, estuaries, marine wetlands, and coastal marine waters [4].

4.4 Sea otters

The sea otters are the littlest marine mammals; they live in the Pacific Ocean (From the coasts of the northern and eastern). The weight of adult sea otters ranged from 14 to 45 kg) (**Figure** 5).

4.5 Polar bears

The polar bears (*Ursus maritimus*) [5] live within the polar circle, encompassing the ocean. They are highly carnivorous animals and the largest extant land carnivore [6, 7]. Adult polar bear weights from 350 to 700 kg (**Figure 6**).

5. Marine conservation

5.1 Why marine conservation is important?

1. Sustaining healthy marine mammal populations is very important in maintaining balance in marine food webs and helping to stay marine ecosystems functioning as they ought to. Additionally, nutrient recycling (reusing nutrients for other styles of ocean production) plays an outsized role all told ocean ecosystems.

2. A healthy ocean regulates climate and reduce temperature change impacts. Ocean currents distribute heat across the world, regulating temperature and

weather. The ocean also absorbs over 90% of the warmth and approximately 30% of CO_2 emissions produced by human activities.

3. As a number of the highest predators of the oceans, marine mammals play a crucial role within the organic phenomenon and help ensure balance within the ocean's ecosystem [7–9].

5.2 How can we protect marine mammals?

- To help in the protection of marine mammals and other protected species the following responsibilities should be taken:
    - Eliminate chemicals and Xinobiotics waste.
    - Take role and responsibility to combat water pollution.
    - Forbidding waste materials (Agriculture and industrial) into marine and water resources.
    - Avoid excessive fish hunting.
- We are requiring the authorities to minimize bycatch in fisheries to make sure our fisheries are sustainable and guarded species are given the most effective chance to recover.
- Global leader in marine mammal conservation and sustainable fisheries, with U.S. fisheries abiding by a number of the world's most robust conservation practices, including measures to cut back marine mammal bycatch—a global threat to several populations of marine mammals.
- Global foundations and the Governorates marine Mammals Protection in response to increasing concerns among scientists and therefore the public that significant declines in some species of marine mammals were caused by human activities. The global and the national policy to:
    - Encourage the public to forestall marine mammal species
    - Encourage the public to forestall population stocks from declining beyond the purpose where they ceased

Unfortunately:

1. The basic information remains poorly known for many species, and not only for those considered Data Deficient, but new technologies are getting down to provide new and better data on both the biology of marine mammals and also the ecology of the oceans.
2. Many marine mammals are still unknown to most of the people and will be forever.
3. Many of those rare species are still unknown to most of the people and will be forever.

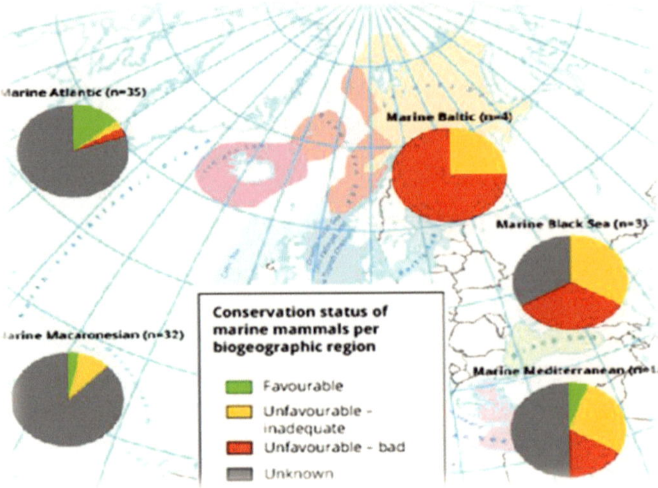

Figure 7.
Drawings are by Sharyn N. Davidson. TM Cox, et al., understanding the impacts of anthropogenic sound on beaked whales. J Cetacean Res Manage 7, 177–187 (2006) [10].

Therefore, our aim emphasizes the importance of understanding the essential biology and ecology of marine mammals to assess the correlates and causes of extinction and to implement science-based conservation (**Figure 7**).

Author details

Hussein Abdelhay Essayed Kaoud
Cairo University, Egypt

*Address all correspondence to: ka-oud@link.net

References

[1] Beerman A, Ashe E, Preedy K, Williams R. Sexual segregation when foraging in an extremely social killer whale population. Behavioral Ecology and Sociobiology. 2016;**70**:189-198. DOI: 10.1007/s00265-015-2038-2

[2] Bengtsson J, Angelstam P, Elmqvist T, Emanuelsson U, Folke C/ Ihse M, et al. Reserves resilience and dynamic landscapes, Ambio 2003;**32**:389-396. DOI: 10.1579/0044-7447-32.6.389

[3] Brakes P, Simmonds MP. Whales and Dolphins: Cognition, Culture, Conservation and Human Perceptions. London: Routledge; 2011

[4] Sinclair ARE, Arcese P, editors. Behavioural ecology for conservation biology: examples from serengeti carnivores. In: Serengeti II: Dynamics, Management, and Conservation of an Ecosystem. Chicago, IL: University of Chicago Press; 2015. pp. 451-472

[5] Ansmann IC, Parra GJ, Chilvers BL, Lanyon JM. Dolphins restructure social system after reduction of commercial fisheries. Animal Behaviour. 2012;**84**:575-581. DOI: 10.1016/j.anbehav.2012.06.009

[6] Stirling I, Lunn NJ, Iacozza J, Elliott C, Obbard M. Polar bear distribution and abundance on the southwestern Hudson Bay coast during the open water season, in relation to population trends and annual ice patterns. Arctic. 2004;**57**:15-26

[7] Stirling I, Smith TG. Implications of warm temperatures and an unusual rain event for the survival of ringed seals on the coast of southeastern Baffin Island. Arctic. 2004;**57**:59-67

[8] Schmittner A. Decline of the marine ecosystem caused by a reduction in the Atlantic overturning circulation. Nature. 2005;**343**:628-633

[9] Northridge SP, Hofman RJ. Marine mammal interactions with fisheries. In: Twiss JR, Reeves RR, editors. Conservation and Management of Marine Mammals. Washington: Smithsonian Institution Press; 1999. pp. 120-155

[10] Davidson SN, Cox TM, et al. Understanding the impacts of anthropogenic sound on beaked whales. Journal of Cetacean Research and Management. 2006;**7**:177-187

Chapter 2

Phylogeny and Population Genetic Structure of Minke Whales Worldwide: A Review of Recent Studies

Luis A. Pastene, Mutsuo Goto, Mioko Taguchi and Yoshihiro Fujise

Abstract

In 1998, two species of minke whales were recognized based on the review of the morphological and genetic information available at that time: the Antarctic minke whale (*Balaenoptera bonaerensis*), which is restricted to the Southern Hemisphere, and the cosmopolitan common minke whale (*Balaenoptera acutorostrata*). Furthermore, three sub-species of the common minke whale were recognized: the North Atlantic (*B. a. acutorostrata*), North Pacific (*B. a. scammoni*) and Southern Hemisphere (*B. a. subsp.*). This chapter reviews the genetic studies on minke whales conducted after 1998. The review is organized by topic, e.g., those studies focused on phylogeny and other matters most relevant for taxonomy, and those focused on population genetic structure within oceanic basins most relevant for conservation and management. On the former topic, the new genetic information, whilst strongly supporting the minke whale taxonomic classification recognized in 1998, also reveals substantial genetic differentiation within the Southern Hemisphere common minke whales, with subsequent taxonomic implications. On the latter topic, results from different analytical procedures have provided information on population identification and structure in the Indo-Pacific sector of the Antarctic and western North Pacific, but they have failed to identify unequivocally any population within the North Atlantic common minke whales.

Keywords: Antarctic minke whale, North Pacific common minke whale, North Atlantic common minke whale, Southern Hemisphere common minke whale, dwarf minke whale, genetics, taxonomy, population structure

1. Introduction

Minke whales are members of the Order Cetacea. They are the smallest species within the suborder Mysticeti (baleen whales), usually not exceeding the 10 m in body length. They are characterized by a sharply pointed head that looks V-shaped when see from above, and they present a sharp longitudinal ridge that runs along the top of the rostrum [1]. Minke whales are the most abundant of the baleen whales and they are hunted in limited numbers by some countries for commercial (Japan and Norway) or aboriginal subsistence (Greenland) purposes.

Until relatively recently, only one species of minke whale was thought to exist: *Balaenoptera acutorostrata*. This even though historical morphological [2–6] and genetics [7–9] data collected from extant populations pointed out to substantial differentiation within the minke whales.

Figure 1.
External morphology of minke whales. From top to bottom: Antarctic minke whale, North Pacific common minke whale, North Atlantic common minke whale and Southern Hemisphere common minke whale (dwarf minke whale).

In 1998, based on a review of both morphological and genetic data, two species of minke whales were recognized, the Antarctic minke whale (*Balaenoptera bonaerensis*), which is restricted to the Southern Hemisphere, and the cosmopolitan common minke whale (*B. acutorostrata*) [10]. Furthermore, three sub-species of the common minke whale were recognized, North Atlantic (*B. a. acutorostrata*), North Pacific (*B. a. scammoni*) and Southern Hemisphere (*B. a. subsp.*) [10]. The common minke whale in the Southern Hemisphere is commonly referred to as the 'dwarf' minke whale [6]. **Figure 1** shows the external morphology of minke whale species and sub-species. As seen in **Figure 1**, the main external morphological character that most readily distinguished the two species is a white flipper patch that is only present in the common minke whale.

Several genetic studies of minke whales have been conducted since the 1998 review. Some studies have focused on phylogenetic issues while others have focused on elucidating population genetic structure in each oceanic basin. This chapter aims to provide a short review of recent genetic studies, outlining the main new findings and implications. After introducing the genetic markers in Section 2, in Section 3, we review the studies that focus primarily on phylogeny and other matters that are relevant to taxonomy and then, in Section 4, we concentrate on the studies on the population genetic structure of each species and sub-species by oceanic basin (Southern Hemisphere, North Atlantic and North Pacific).

Both information on taxonomy and population identification and structure of minke whales are important and necessary for effective decision-making about conservation and sustainable use of the species.

2. Genetic markers

Two main genetic markers have been used in recent genetic analyses of minke whales, mitochondrial DNA (mtDNA) control region sequences and microsatellite DNA (msDNA, a nuclear marker) genotypes, which are briefly explained here based on [11].

The mitochondrial genome is a circular, double-stranded molecule ranging in size from 16,500 to 17,600 base-pairs (bp) in cetaceans. The main features of mtDNA are (a) maternal inheritance, (b) no recombination during reproduction and (c) it is haploid. Features (a) and (c) mean that the effective population size for the mtDNA genome is ¼ of that for nuclear markers. Sequence changes in animal mitochondrial genomes are of four types: sequence arrangements; additions; deletions; and nucleotide substitutions. The substitution rate is not constant across the mitochondrial genome. The most variable part is where replication begins (the 'control region'). The control region is the only major non-coding region in the mitochondrial genome. In whales, its length is approximately 1000 bp. In most studies on minke whales, the sequence of the first 300-500 bp in the control region is determined, which is the most variable part.

MsDNA or simple tandem repeats (STRs) are segments of non-coding nuclear DNA containing a varying number (different alleles) of tandem repeats of short sequences of less than six nucleotides. As a nuclear marker, they are diploid with recombination during reproduction. They are abundant and widely distributed throughout the mammalian genome. MsDNA is highly variable, presenting a large number of alleles at each locus, selectively neutral, inherited in standard Mendelian fashion and allelically codominant. MsDNA generally evolves by changes in the number of repeats, i.e., in the length of the repetitive region. MsDNA alleles can be distinguished by differences in the length of the repetitive region. They predominantly mutate by insertion or deletion of repeats. In most studies on minke whales, a set of approximately 12–16 msDNA loci are used.

Most of the recent genetic studies on minke whales have made combined use of these two genetic markers, which presents several advantages. Some of the genetic criteria for taxonomic definition require results of both markers (see below). Different species of large whales can produce hybrid whales and such cases can be detected by the combined use of mtDNA and msDNA. In studies on population identification and structure, parallel analyses of Mendelian and maternally inherited loci are particularly important. Some species may display maternally directed phylopatry. In such cases, genetic differences can be found for the mtDNA but not for msDNA. The use of msDNA in addition to mtDNA allows for an investigation of kinship, which is important information for the interpretation of population structure.

Details of laboratory procedures for mtDNA and msDNA in minke whales can be found in [12].

3. Phylogenetic and other studies relevant for taxonomy

Several genetic studies addressing phylogenetic and other aspects relevant for taxonomy were conducted after the 1998 review in [10]. All those studies used samples from minke whale worldwide [13–18]. Oceanic basins covered by the genetic sampling in recent studies are shown in **Figure 2**.

A brief description and main findings of these studies are presented below. Several phylogenetic inference methods were used to evaluate observed heritable traits, such as mtDNA sequences, under a specified model of the evolution of the traits. Taxonomic classification is now usually based on phylogenetic data. Details of the phylogenetic inference methods are not given here however relevant bibliographic references on the methods are provided for interested readers in the sections below.

3.1 Speciation and divergence time

The focus of the first post-1998 study involving minke whales was a case study to investigate the radiation and speciation of pelagic organisms during the period of global warming [13]. The study was based on mtDNA control region sequences (340 bp) in samples of Antarctic minke whales (n = 180), North Atlantic (n = 102) and North Pacific (n = 161) common minke whales and Southern Hemisphere common or dwarf minke whales (n = 23 from the western South Atlantic, WSA and western South Pacific, WSP). A total of 187 haplotypes (unique sequences) were determined. The genealogical relationship among a sub-set of 60 haplotypes was estimated using the NUCML program in the MOLPHY computer package [19], the BASEML program in the PAML computer package [20] and the TREE-PUZZLE program of the quartet-puzzling (QP) method [21]. Divergence time was estimated by applying a molecular clock model using a calibration point that minke whales and the gray whales (*Eschrichtius robustus*) separated 20 million years ago (Ma) [22].

The study provided evidence for phylogenetic differentiation not only between the two species of minke whales but also among North Atlantic, North Pacific and Southern Hemisphere common minke whales. The study estimated that the two species of minke whales diverged in the Southern Hemisphere less than 5 Ma, and that the current sub-species of the common minke whales diverged after the Pliocene some 1.5 Ma.

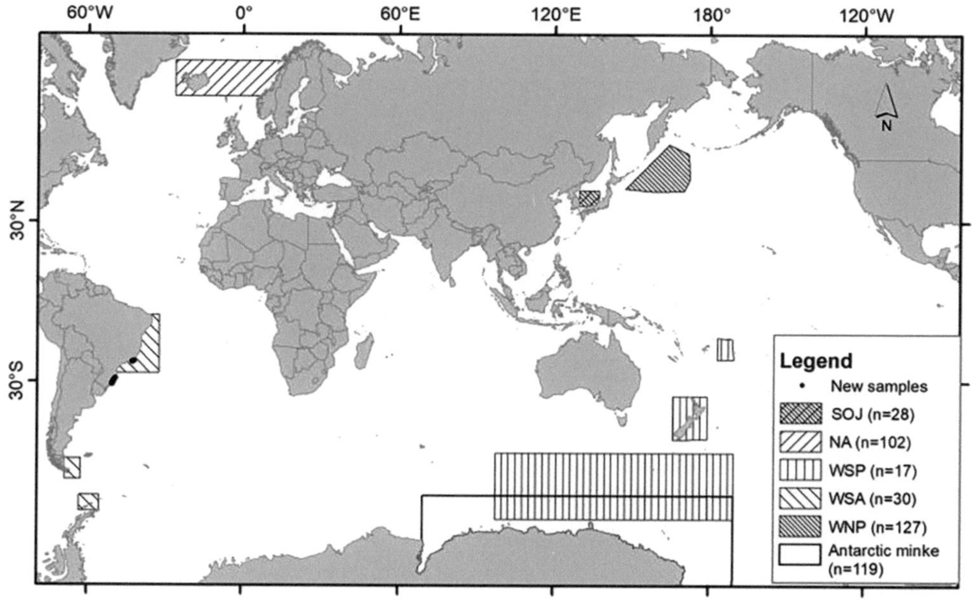

Figure 2.
Oceanic basins covered by the genetic sampling for the phylogenetic and other studies relevant for the taxonomy of minke whales. SOJ = Sea of Japan, NA = North Atlantic, WSP=Western South Pacific, WSA = Western South Atlantic, WNP=Western North Pacific, Antarctic minke whale = Antarctic Ocean (modified from [18]).

Based on their analysis, the authors hypothesized that prolonged periods of global warming facilitate speciation in pelagic marine species that depend on upwelling [13].

3.2 Phylogenetic analyses

Three relevant studies are described here [14, 16, 18]. The first study [14] used mtDNA control region sequences (327 bp) and a similar sample set of the previous study [13] but this time the study was focused to elucidate the population genetic structure of the Southern Hemisphere common minke whales using samples from WSA (n = 12) and WSP (n = 17) (**Figure 3**).

The genealogy of the mtDNA haplotypes was estimated using the neighbor-joining method (NJ) [23], minimum evolution (ME) [24], maximum likelihood (ML) [25] and maximum parsimony (MP) [26]. To evaluate the relative effects of divergence and migration between WSA and WSP whales, the approach in [27] modified for a finite mutation level [28] was used. Phylogenetic inferences derived from these methods were consistent, and similar to the inferences obtained in a previous study [13]. WSA common minke whale haplotypes (except one), clustered in a single clade, which nested within the North Atlantic common minke whale clade. On the other hand, WSP common minke whale haplotypes clustered in a different clade. The study showed that haplotypes from the WSA whales share more recent common ancestors with the North Atlantic minke whales than they do with the WSP minke whales. The analysis suggested a very low number of migrants by generation between WSA and WSP, which suggests that the WSA single haplotype in the WSP clade was unlikely to be a result of migration but rather due to incomplete lineage sorting [14].

The most recent genetic analysis on minke whales worldwide [18] was based on mtDNA control region sequences (313 bp) and msDNA (11 loci). The sample

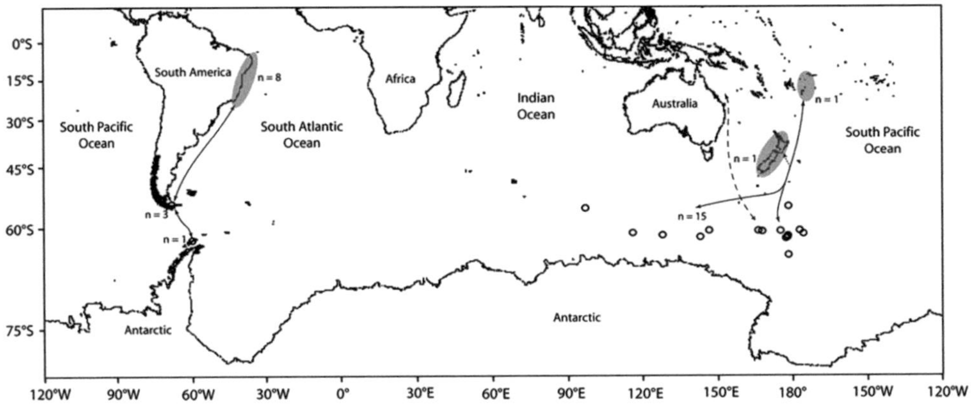

Figure 3.
The geographic position of Southern Hemisphere common minke whales (dwarf minke whales) samples used in [14]. Solid and dashed lines indicate possible migratory routes and possible connections, respectively (modified from [14]).

set for the mtDNA analysis was similar to those in the previous studies [13, 14] but the samples of the Southern Hemisphere common minke whales were increased (WSP, n = 17; WSA, n = 30), and msDNA was used in addition to mtDNA. A total of 148 haplotypes were determined. The genealogy of the mtDNA haplotypes was estimated using several methods including NJ, ML and Bayesian inferences (BI) [29]. The three methods provided similar results, and they were consistent with previous phylogenetic inferences [13, 14]. Results from the BI method are shown in **Figure 4**. This figure shows two main clades, one corresponding to Antarctic minke whales and the other to common minke whales. Furthermore, within the common minke whales clade, North Pacific, North Atlantic and Southern Hemisphere common minke whales clustered in different sub-clades.

Figure 4 shows that WSA and WSP common minke whales in the Southern Hemisphere clustered in different sub-clades (except the single WSA haplotype mentioned previously that clustered within the WSP sub-clade), and that the WSA haplotypes fell with the North Atlantic sub-clade.

This study estimated the net nucleotide substitutions (d_A) [30] between species and sub-species of minke whales. The d_A between the Antarctic and common minke whales was high (0.08 in average). The value among common minke whales from different oceanic basins averaged 0.026. The d_A between Southern Hemisphere WSP and WSA was 0.027 and that between the Sea of Japan and western North Pacific was 0.007 [18].

The msDNA analysis in [18] involved samples from three localities only (unfortunately, no samples from the North Atlantic common minke whales were considered): North Pacific and Southern Hemisphere (WSA and WSP) common minke whales. The pattern of msDNA differentiation was investigated by two indices, F_{ST} [31] and D_{SW} [32]. All pairwise comparisons among North Pacific, WSA and WSP yielded statistically significant differences and the values estimated between WSA and WSP were smaller than the values between each of these populations and North Pacific common minke whales. Therefore, North Pacific, Southern Hemisphere WSA and WSP not only were separated phylogenetically in their mtDNA but they differed significantly in their msDNA as well.

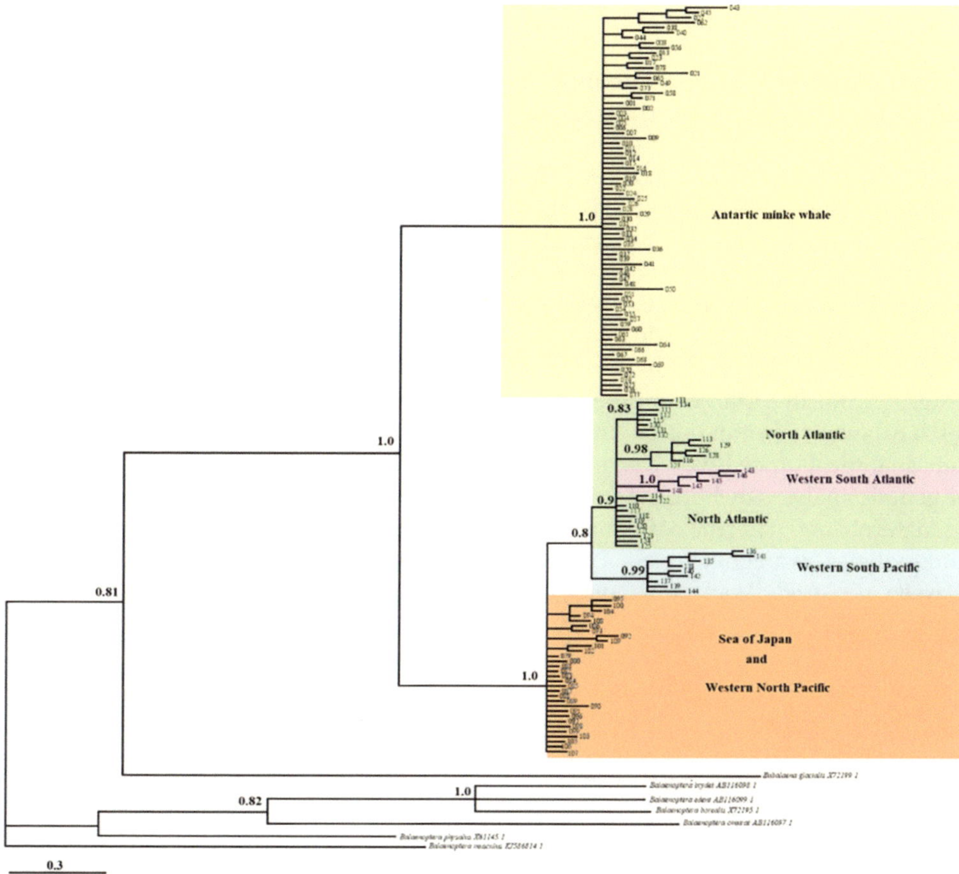

Figure 4.

Bayesian phylogenetic tree of minke whale mtDNA haplotypes. Values indicate support for each node according to the maximum posterior probabilities>80%. Scale bar represents substitutions per nucleotide site. NA = North Atlantic; WSA: Western South Atlantic; WSP = Western South Pacific; SOJ = Sea of Japan; WNP = Western North Pacific (modified from [18]).

Although, the third study was focused to investigate hybrids between the two species of minke whales [16], it also provided information on genetic differentiation between the Antarctic and common minke whales species as well among common minke whales from different oceanic basins. The study was based on mtDNA control region sequences (287 bp) and msDNA (11 loci), and samples from the Antarctic minke whale (n = 91), North Atlantic (n = 91), North Pacific (n = 95) and Southern Hemisphere (WSP) (n = 9) common minke whales. The genealogy of the mtDNA haplotype was estimated using the NJ method and the inferences obtained were similar to the other studies [13–14, 18]. The msDNA F_{ST} estimates were calculated and Bayesian cluster analysis was also performed using the program *STRUCTURE* [33]. Pairwise F_{ST} estimates revealed that the Antarctic minke whales, North Atlantic, North Pacific and Southern Hemisphere (WSP) common minke whales were genetically distinct from each other. The Bayesian cluster analysis supported the F_{ST} results, showing large genetic differences between the Antarctic and common minke whales as well among common minke whales from North Atlantic, North Pacific and Southern Hemisphere (WSP) [16].

3.3 Hybridization in minke whales

A genetic study based on both mtDNA (287 bp) control region sequences and msDNA (13 loci) reported the migration of an Antarctic minke whale into the Arctic Northeast Atlantic in 1996 [15]. The same study reported the occurrence of a hybrid whale in the North Atlantic in 2007. The analytical procedures for the identification of the hybrid involved the use of the Bayesian cluster analysis *STRUCTURE* and genetic assignment conducted in the program GeneClass2 [34]. The latter used a genetic baseline consisting of the three minke whale species and sub-species which had a large sample size (Southern Hemisphere common minke whales were excluded due to their small sample size), in addition to three sets of hybrids produced in the program HYBRIDLAB1.0 [35]. The 2007 hybrid was demonstrated to consist of a maternal contribution from an Antarctic minke whale and most likely paternal contribution from the North Atlantic common minke whale. Another case of a hybrid was identified using the same analytical procedures. It was a pregnant female captured in 2010 [16]. In this case, the genetic analyses by both markers confirmed that the mother was a hybrid displaying maternal and paternal contribution from North Atlantic common and Antarctic minke whales, respectively [16]. This study demonstrated for the first time, that hybrids between minke whale species may be fertile, and that they can back-cross.

3.4 Implications for taxonomy and suggestions for future works

Taxonomic definitions are associated with the term Evolutionary Significant Unit (ESU) [36, 37], defined in [37] as 'ESUs should be reciprocally monophyletic for mitochondrial DNA alleles and show significant divergence of allele frequencies at nuclear loci'. However, other authors have argued that the definition of ESUs should incorporate ecological data in addition to data on genetic variation of adaptive significance [38]. An example of ecological data could be discrete prey preferences of sympatric individuals. Other authors suggest the use of d_A values based on mtDNA: a review of analytical approaches for recognition of populations, sub-species and species based on mtDNA sequences suggested that species generally exhibit values of d_A greater than 0.02 and populations values less than 0.004 [39], and see also [18].

Considering these criteria, the post-1998 genetic results (with larger sample sizes and wider geographical range), strongly support the division of Antarctic and common minke whales as different species [10]. They clearly match the ESU definition (based on different phylogenetic inference methods), and the average estimated d_A between the Antarctic and common minke whales from different oceanic basins was estimated at 0.08.

Within the common minke whales, the North Pacific and Southern Hemisphere (WSP) match the ESU criterion. Their average d_A with common minke whales from other oceanic basins averaged 0.02 [18]. Then the status of sub-species is appropriated for North Pacific and Southern Hemisphere (WSP) common minke whales.

The case of the North Atlantic and Southern Hemisphere (WSA) common minke whales is more complex. This is because some of the mtDNA phylogenetic analyses showed haplotypes of common minke whales from WSA clustering within the North Atlantic common minke whale clade, therefore not matching the reciprocally monophyletic for mitochondrial DNA definition of ESU, although the status of sub-species is appropriate based on the d_A criterion. Therefore, while both Southern Hemisphere common minke whales (WSP and WSA) are clearly separated from North Pacific common minke whales matching all criteria for sub-species, the relationship between

WSA and North Atlantic common minke whales requires further investigation including additional genetic analyses based on larger samples from WSP and WSA using both mitochondrial and nuclear markers. In addition, genetic analyses of Southern Hemisphere common minke whales from other unstudied localities, e.g., the Western Indian Ocean [6], are required to elucidate further the phylogenetic relationship among Southern Hemisphere and North Atlantic common minke whales.

Finally, and following the criteria above, whales from the Sea of Japan and western North Pacific should be considered as populations of the North Pacific common minke whale.

The cases of hybridization between minke whale species and the study showing that such hybrids may be fertile, and that they can back-cross have some relevance to the taxonomy of minke whales. As noted in [16], it is not possible to resolve whether the observed migration of Antarctic minke whales to the Arctic, and hybridization between Antarctic minke whales and North Atlantic common minke whales are (a) random events that have occurred over a long period of time; (b) the result of a low number of Antarctic minke whales migrating from the Antarctic to the Arctic in the 1990s; or (c) represent a trend that is increasing in frequency. The authors in [16] further argued that the lack of hybrids in the large (n > 15000) Japanese genetic data sets infers that such events are not frequent. Unless the frequency of reproductive contact increases significantly in the future, the separation of the Antarctic minke whale and the North Atlantic common minke whale should not be challenged [16].

In summary, the recent genetic studies provide support for the classification recognized in the 1998 review [10] for two species, the Antarctic and the common minke whale, and at least three sub-species of the latter. Furthermore, these studies suggest a phylogenetic separation between Southern Hemisphere common minke whales from Western South Pacific and Western South Atlantic. Whales from these two localities differed significantly in mtDNA haplotype and msDNA allele frequencies. Phylogenetic analyses showed that haplotypes from the WSA whales share a more recent common ancestor with the North Atlantic common minke whales than they do with the WSP common minke whales.

4. Studies on population genetic structure in each oceanic basin

Minke whales were hunted commercially or under special permit in the Southern Hemisphere until the 2018/19 austral summer season, and they are hunted currently for limited numbers in the North Atlantic (commercial and aboriginal subsistence purposes), and western North Pacific (commercial purposes). Identification of populations within species and sub-species in each oceanic basin, therefore, is very important for conservation and management purposes. This is because different populations of the same species or sub-species may respond in different ways to levels of direct removals (e.g., catches, bycatches) and other types of environmental stress (e.g., habitat degradation) [18]. Population dynamics modeling is used to investigate the effect of different management strategies and environmental stressors at the population level. However, the identification of populations is not a trivial issue.

In each of the relevant oceanic basins, Southern Hemisphere, North Atlantic and North Pacific, minke whales are believed, like most baleen whale species, to undertake seasonal migrations between feeding grounds in higher latitudes in summer and breeding grounds in lower latitudes in the tropical or temperate regions in winter.

However, there are few direct observations of this linkage, and information of minke whale breeding grounds in low latitudes is poor. Ideally, genetic analyses on population identification should be carried out based on samples collected in breeding grounds. However, all genetic analyses on minke whale population identification have been based on samples collected in feeding grounds and migratory corridors, where different populations may mix spatially and/or temporally.

The International Whaling Commission (IWC) has defined areas for the management (i.e., the setting of catch limits) of minke whales in each oceanic basin based upon a variety of data types, genetic and non-genetic (e.g., see [40]) since the earliest days of management, often based upon limited information or analogy. Most recent studies have focused on the correspondence of the set management boundaries with the available genetic information and revising the boundaries as appropriate to ensure that over-exploitation does not occur. The primary management tool used by the IWC Scientific Committee to provide advice on commercial whaling catch limits is known as the Revised Management Procedure or RMP that focusses on providing robust management advice in the light of inevitable scientific uncertainty (e.g., [41]). Uncertainty in stock structure is one of the most influential in terms of providing robust advice. The philosophy adopted under the RMP (and the sister approach for aboriginal subsistence whaling known as the AWMP or Aboriginal Subsistence Whaling Management Procedure) with respect to stock structure is that it is not often, if ever, possible to arrive at only one plausible stock structure hypothesis from the available data. Rather than in the past when the 'best' hypothesis (and boundaries) was determined and then fixed management boundaries for the 'unit-to-conserve' (usually a population) chosen, the RMP approach says that catch limits must be set that are robust to all plausible hypotheses and that these hypotheses should be regularly reviewed in the light of new data. Of course, deciding what comprises 'plausible' is a complex and difficult issue and one which has driven much of the work described below, especially for the North Pacific common minke whale.

In this section, the most recent genetic analyses on population identification and structure in minke whales are reviewed for each species and sub-species in each relevant oceanic basin.

The method most often used for the identification of populations within an oceanic basin was hypothesis testing under the null hypothesis of panmixia. Under this method, mtDNA haplotype and/or msDNA allele frequencies between two geographically grouped samples are compared using several statistical tests. More recently, spatially explicit clustering approaches, for example, sPCA, GENELAND, TESS and BAPS have been used to investigate population identification and structure.

Details of the statistical tests and clustering approaches are not given here however relevant references on the methods are provided for interested readers in the sections below.

4.1 Antarctic minke whales

The IWC's management areas for baleen whales (excluding the Bryde's whale *Balaenoptera edeni*) are shown in **Figure 5**. These management areas were used during the former commercial whaling of Antarctic minke whales but were based upon information from other baleen whales, notably blue (*B. musculus*), fin (*B. physalus*) and humpback (*Megaptera novaeangliae*) whale catch distributions and mark-recapture records. Most of the recent genetic studies have been focused in the Indian and Pacific sectors of the Antarctic (mainly Areas IV and V in **Figure 5**) where a large number of genetic samples were available from the Japanese Whale Research Program

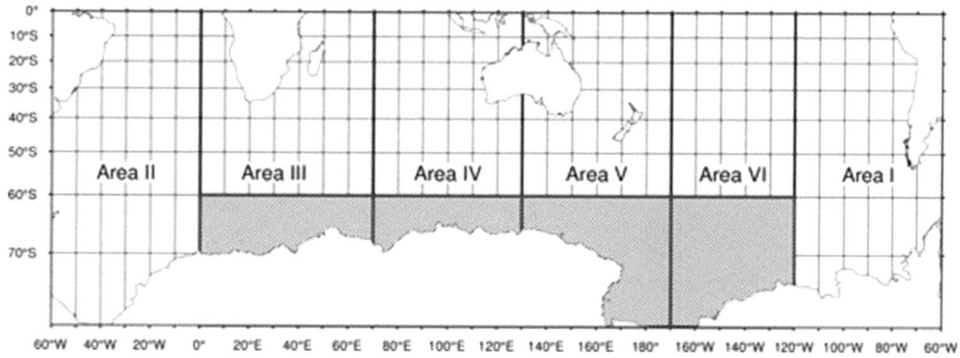

Figure 5.
Management areas defined by the International Whaling Commission for the management of baleen whales (except the Bryde's whale) in the Southern Hemisphere. These areas were used for the management of the Antarctic minke whale in the period of commercial whaling, which was stopped in the 1986/87 austral summer season. Most of the recent genetic studies on population structure have been conducted in the shaded area.

under Special Permit in the Antarctic, Phases I and II (JARPA and JARPAII). Surveys of these research programs were conducted systematically in the Indo Pacific sector of the Antarctic in summer from 1987/88 to 2014/15.

There are no genetic samples from Antarctic minke whales from low latitude regions of the eastern Indian Ocean and western South Pacific where breeding grounds of this species in this region are assumed to occur. The most recent genetic studies were based therefore on samples collected by the JARPA and JARPAII programs in the Antarctic feeding grounds of Areas III east, IV, V and VI west. Those studies were summarized in [42], and the most relevant aspects are highlighted here.

Previous morphometric, biological and genetic studies based on mtDNA and msDNA led to the conclusion that Antarctic minke whales in the feeding grounds between Areas III east and VI west do not comprise a single population [43]. The most recent genetic study used mtDNA control region sequences (340 bp) and msDNA (12 loci) [12] to examine a total of 2254 samples in the Indo-Pacific sector of the Antarctic: Area III east = 564; Area IV west = 734, Area IV east = 74, Area VE east = 478, Area VI west = 404. The samples were obtained in the Southern Hemisphere summer season in different years. The degree of spatial and temporal divergence was estimated via the F_{ST} and by the randomized chi-square Test of Independence [44]. Results of the heterogeneity tests for both markers showed statistically significant genetic differences between whales in the most distant sectors, western (35°- 130°E) and eastern (165°E - 145°W) (see **Figure 5**), confirming that different populations inhabit the Indian and Pacific sectors of the Antarctic. A simulation study on the dynamics of the species showed that both populations had a soft boundary in the sector 100°-165°E [45].

The main conclusion of the studies was the existence of at least two populations in the feeding grounds of the Indo-Pacific sector of the Antarctic and a transition area in the region around 100°-165°E, across which there is an as yet undetermined level and range of mixing (**Figure 6**). The following names were proposed for these populations: Eastern Indian Ocean Population (I-Population) and Western South Pacific Ocean Population (P-Population) [42].

A recent study described a paternity method based on msDNA (12 loci) to estimate the abundance of mature male Antarctic minke whales in the Indo Pacific sector of the Antarctic using a maximum likelihood approach [46]. Results for the geographical distribution of mother/fetus-father pairs were generally consistent with

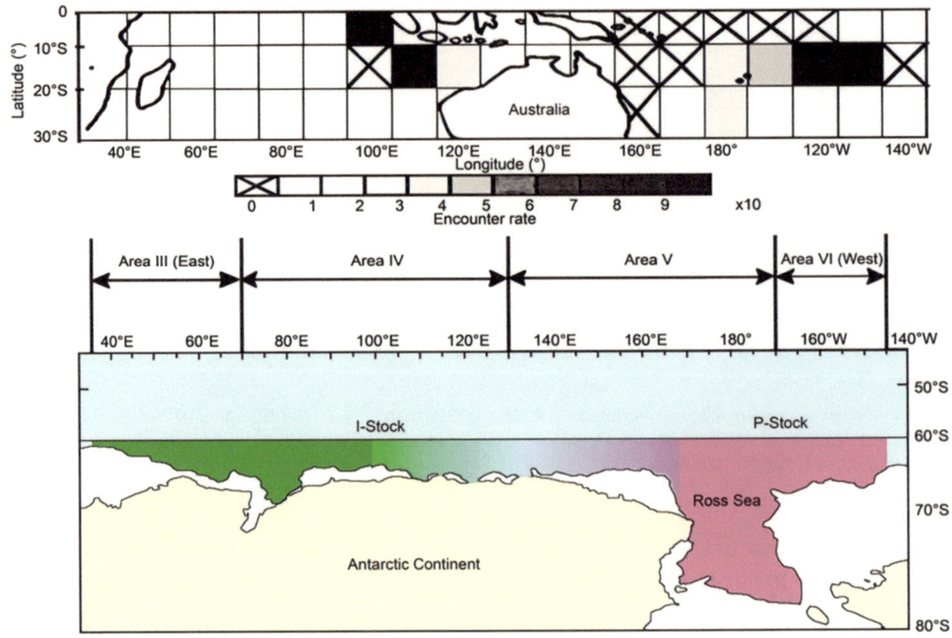

Figure 6.
The current hypothesis of population structure of the Antarctic minke whale. At least two populations occur in the Indo-Pacific sector of the Antarctic covered by the surveys of the JARPA/JARPAII, which mix in a transition area, whose position and extension varies by year and sex. These populations are possibly related to breeding grounds in lower latitudes evidenced by high-density areas suggested by sighting surveys (upper part of figures) (after [42]).

the hypothesis of separate I- and P- Populations because eight of 10 pairs were found in the expected areas of distribution of either population. Only two pairs were found in distant areas.

The genetic studies showed no concordance between the geographic boundaries of the IWC management Areas and the geographical distribution of the I- and P- populations suggested by the genetic analyses.

4.2 North Atlantic common minke whale

The IWC's management areas for North Atlantic common minke whales are shown in **Figure 7**. In this section, the most recent genetic studies on population structure are summarized [47–49]. These studies were focused on examining the biological validity of the management areas in **Figure 7**.

The first study reviewed here [47] was based on genetic samples (n = 306) collected throughout the North Atlantic (see **Table 1**). Samples were collected in spring-summer over several years. The genetic markers used were mtDNA control region (500 bp) and msDNA (16 loci). The analytical procedures used for mtDNA were the F_{ST} for haplotype frequencies and the PHI_{ST} [50]. MsDNA variation was analyzed by testing for homogeneity of allele frequencies among populations using GENEPOP [51] and F_{ST}. Based on the combination of several approaches the authors suggested the existence of four genetically differentiated populations: (1) West Greenland; (2) Central North Atlantic-East Greenland-Jan Mayen; (3) North East Atlantic including Svalbard, the

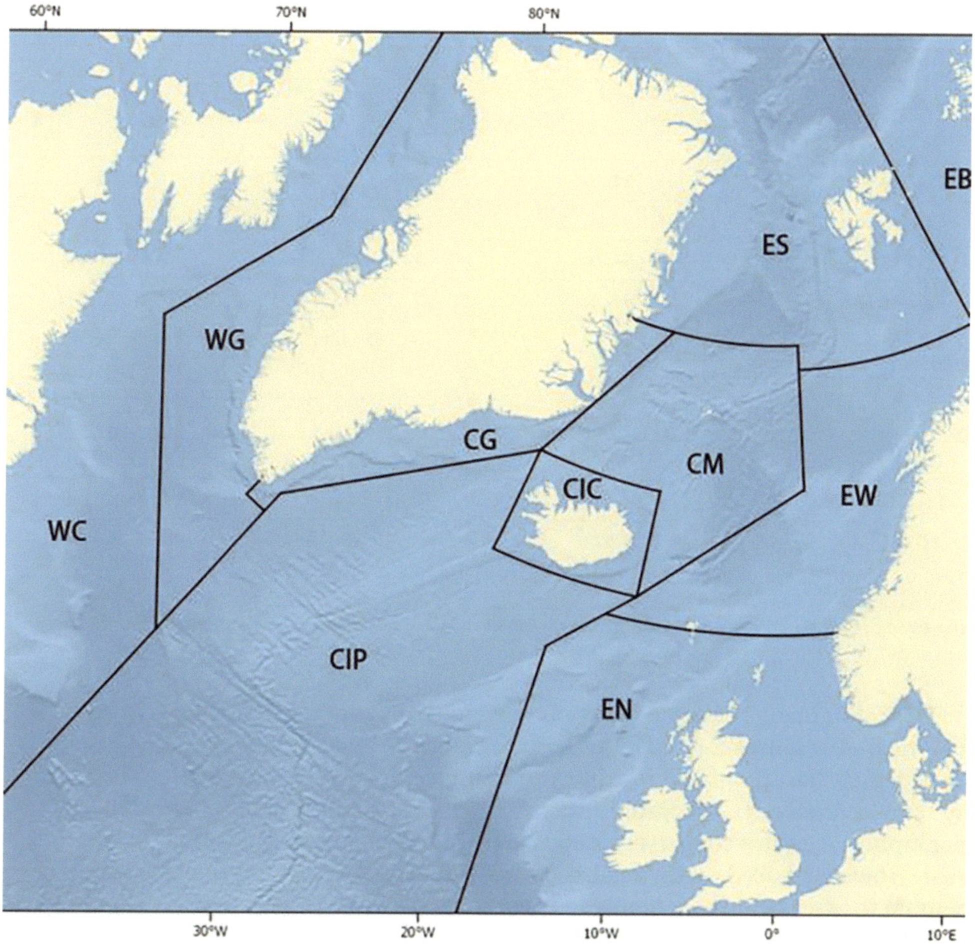

Figure 7.
Management subareas used by the International Whaling Commission for the management of commercial and aboriginal subsistence whaling of North Atlantic common minke whales. Sub-areas prefixed by W represent the western North Atlantic, sub-areas prefixed by C represent the central NorthAtlantic and sub-areas prefixed by E represent the eastern North Atlantic. Management subarea EC mentioned in the main text merged into a single EW subarea.

Barents Sea and north western Norway, and (4) the North Sea. Unlike the other areas, there was a lack of inter-annual variation in West Greenland. The authors postulated that each population evolved in response to regional differences in ecological conditions, namely oceanography, ice cover, prey type and prey availability [47].

The second study [48] was based on smaller sample size (n = 202) but again throughout the North Atlantic (see **Table 1**). Samples were collected mainly in spring-summer over several years. The genetic markers used were mtDNA control region sequences (345 bp) and msDNA (10 loci). The relevant analytical procedures to investigate population structure based on msDNA were the F_{ST} and Rho_{ST} [52]. Also, the study estimated the most probable number of putative populations (K) using *STRUCTURE*. To facilitate the interpretation of the *STRUCTURE* output, a measure based on the second order rate of change of the likelihood function with respect to K was plotted [53]. The F_{ST} and Rho_{ST} were calculated for the population suggested in

Management Area	Sample size study [47]	Sample size study [48]	Sample size study [49]
Western NA	**166**	**51**	**0**
WG	166	36	
WC		15	
Central NA	**54**	**17**	**48**
CIC			
CG + CM	54	17 (CM only)	48 (CM only)
Eastern NA	**86**	**131**	**2596**
ES, EB, EC	63	48 (ES only)	1583 (ES + EB only)
EW			1013
EN	23	83	
Other	**0**	**3**	**0**
Spain		3	
TOTAL	**306**	**202**	**2664**

Table 1.
Summary of sample sizes by North Atlantic management subareas in the three studies referred to in the text.

STRUCTURE using the same methods used for the geographical comparisons. The analytical procedures for mtDNA were the same F_{ST} and PHI_{ST} used in the previous study, which was calculated for the populations inferred from the *STRUCTURE* in the same way as for the geographic comparisons. The study found no evidence of geographic structure comparing putative populations in recognized management areas. However, based on the results of individual genotypes and likelihood assignment methods, the authors identified two putative 'cryptic' populations (populations exhibiting some level of genetic structure, which cannot be explained by past or current barriers to dispersal alone) distributed across the North Atlantic in similar proportion in different regions. They suggested that common minke whales range extensively across the North Atlantic seasonally, but segregate to some extent on at least two breeding grounds [48].

The third study [49] was based on much larger sample size (n = 2664) but primarily from the Eastern North Atlantic (**Table 1**). The genetic markers used were mtDNA control region sequences (331 bp) and msDNA (10 loci). The study used several analytical procedures to investigate population structure based on msDNA including *STRUCTURE*, BAPS (Bayesian Analysis of Population Structure) [54] and traditional F_{ST} and R_{ST} [55]. Genetic differentiation among management areas per year, and the level of temporal population genetic differentiation were tested using the Analysis of Molecular Variance (AMOVA) [56]. The possibility of cryptic populations suggested in the previous study [48] was investigated using *STRUCTURE* and two different outgroups. For mtDNA, the relevant analyses on population structure were based on AMOVA. The authors summarized their findings as follows: no spatial or temporal genetic differentiation was observed for either class of genetic marker; mtDNA identified three distinct lineages without any underlying geographical pattern; nuclear markers showed evidence of a single panmictic population in the eastern North Atlantic. Results of additional simulation analyses suggested that clustering methods may spuriously reveal cryptic genetic structure [49].

4.3 North Pacific common minke whale

The IWC's management sub-areas for North Pacific common minke whales are shown in **Figure 8**. At least two populations of the common minke whales have been historically recognized in the western North Pacific, (1) the Okhotsk Sea-West Pacific (known in IWC literature as the O-stock) and (2) the Sea of Japan-Yellow Sea-East China Sea (known as the J-stock). There are morphological and reproductive [57, 58] as well genetic [59, 60] characters differentiating these two populations.

Recent genetic work has focused on refining this two-population hypothesis as well as investigating whether additional structure exists within the J- and O-stocks. Studies have been based on samples collected mainly during the Japanese Whale Research Programs under Special Permit in the western North Pacific, Phases I and II (JARPN and JARPNII) and bycatches along the Japanese coast. Surveys of these research programs were conducted systematically in the western North Pacific in spring-summer from 1994 to 2016. **Table 2** summarizes the number of samples used in recent studies, by subarea.

Individual probability assignment to either J- or O-stocks was made possible by the use of *STRUCTURE* in a study that examined 4275 samples obtained from JARPN/JARPNII and by-catches in the subareas shown in **Figure 8** and **Table 2**, using mtDNA control region sequences (487 bp) and msDNA (16 loci) [61]. Statistical tests were conducted to investigate deviations from expected Hardy–Weinberg genotypic proportions and *STRUCTURE* was used to determine *K*, the most likely number of genetically distinct populations present in the samples. Regarding mtDNA, the genealogy of haplotypes was estimated using the neighbor-joining method. Twelve of the 16 msDNA loci showed significant deviation from the expected Hardy–Weinberg

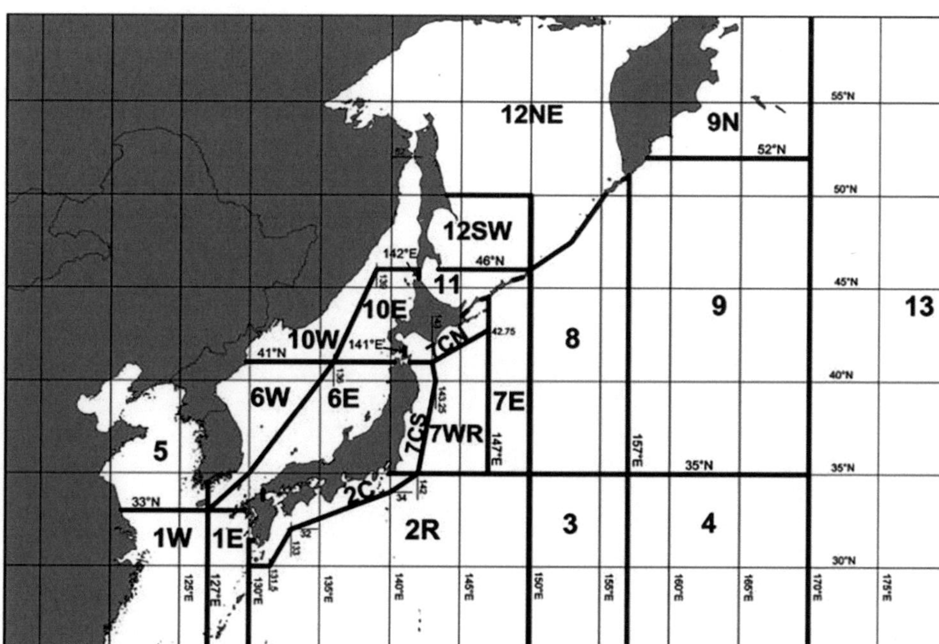

Figure 8.
Management subareas defined by the International Whaling Commission for the management of the North Pacific common minke whales.

Management Area	Sample size study [61]	Sample size O-stock study [62]
2C	487	
6E	717	
10E	13	
11	129	48
7CN	1066	739
7CS	921	439
7E	49	45
7WR	100	89
8	252	223
9	541	487
TOTAL	**4275**	**2070**

Table 2.
Summary of sample sizes by North Pacific management subareas used in recent studies referred to in the text.

genotypic proportions. The inbreeding coefficients were all positive suggesting a homozygote excess. This deviation suggested the existence of individuals from multiple populations in the sample set. The *STRUCTURE* analysis presented the highest likelihood probability at $K = 2$. These results indicated that the samples came from two genetically distinct populations, the J- and O-stocks. **Figure 9** shows the distribution of J and O-stock individuals by sub-area. Almost all the individuals from the Sea of Japan (sub-areas 6E, 10E) were assigned to J-stock, whereas almost all individuals from the offshore North Pacific (east of area 7WR) were assigned to O-stock. Intermediate areas (7CN, 7CS, 11) contained individuals from both stocks. Area 2C on the Pacific side of Japan is mainly occupied by the J-stock individuals.

Figure 10 shows the temporal distribution of the J- and O-stock individuals on the Pacific side of Japan (2C, 7CN and 7CS) expressed as a three-month moving average. In sub-area 2C, J-stock animals are predominant throughout the year. In sub-areas

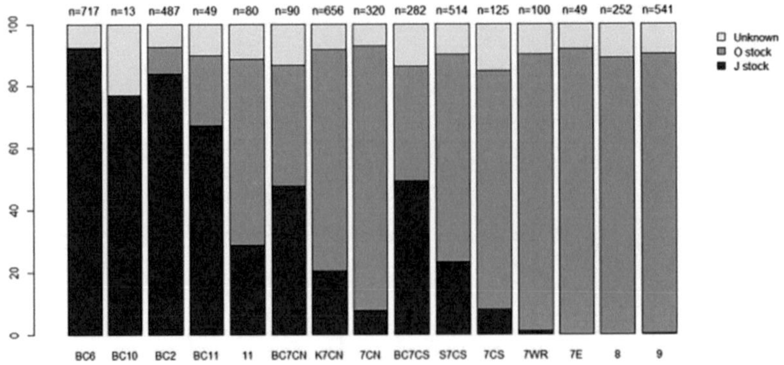

Figure 9.
Spatial occurrence of O- and J-stocks in management sub-areas around Japan (see **Figure 8***). BC2, BC6, BC7CS, BC7CN, BC10, BC11 = bycatches from the respective areas; K7CN = coastal JARPN/JARPNII surveys at Kushiro; S7CS = coastal JARPN/JARPNII surveys at Sanriku; 7CS, 7CN, 7WR, 7E, 8, 9 and 11 = offshore JARPN/JARPNII surveys. Sample sizes are at the top of each bar. 'Unknown' refers to individuals that could not be assigned to either stock by STRUCTURE (after [61]).*

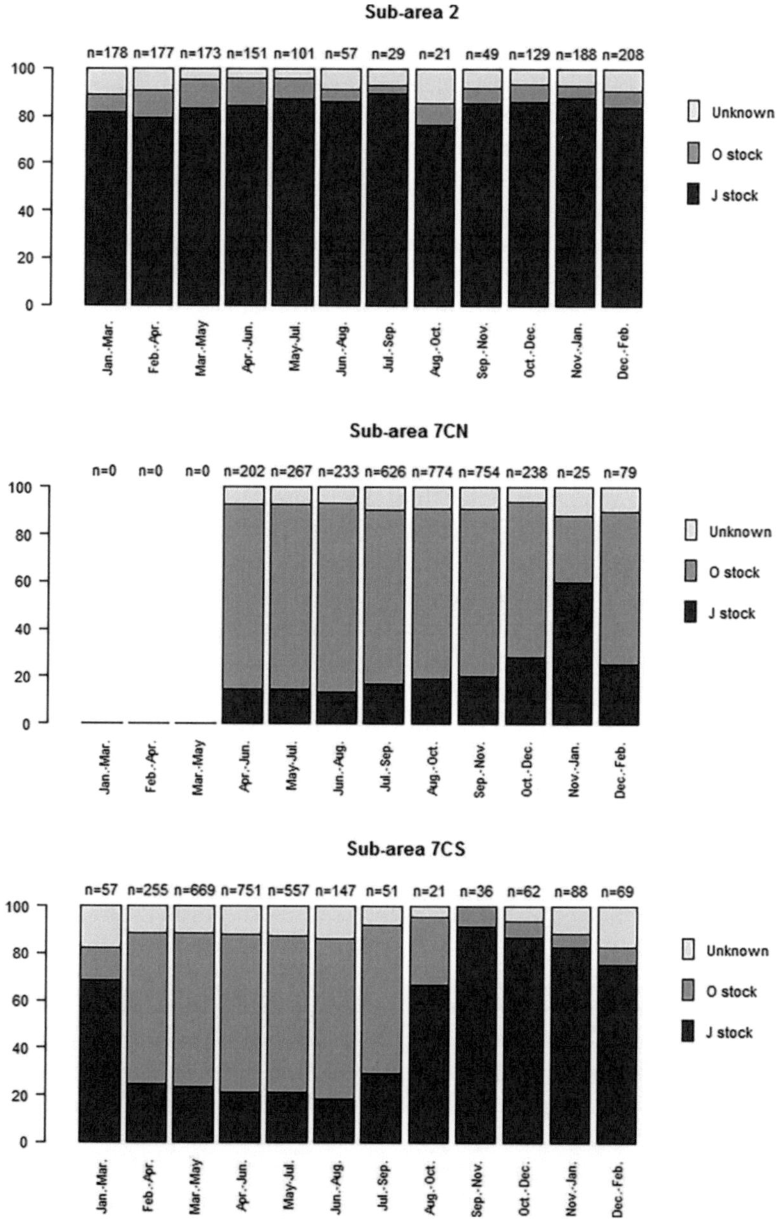

Figure 10.
Monthly occurrence of O- and J-stocks in areas 2C, 7CS and 7CN. Each bar is expressed as three-month moving average. Sample sizes are on the top of each bar. The sampling years in area 2C was 2001–2014; in areas 7CN and 7CS was 1994–2014. 'Unknown' refers to individuals that could not be assigned to either stock by STRUCTURE (after [61]).

7CS and 7CN, the proportion of the J-stock increases in autumn/winter and decreases in spring/summer – the reverse is true for O-stock animals.

The phylogenetic tree of haplotypes showed no population-specific clade although most of the individuals assigned to the J-stock shared the same clade. Most of the individuals assigned to the O-stock shared clades where the J-stock individuals were less frequent [61].

A subsequent study investigated the possibility of additional structure within O-stock based on mtDNA control region sequences (487 bp) and msDNA (16 loci) [62]. The sample size of the O-Stock for the different subareas shown in **Figure 8** was 2070 (**Table 2**). The methods used for investigating structure based on msDNA data were the probability test [63] and the discriminant analysis of principal component (DAPC) approach [64]; for the latter analysis, both J- and O-stock assigned individuals were used. For mtDNA, heterogeneity tests in haplotype frequencies among the samples were conducted using both the chi-square test of independence and conventional F_{ST}. Results based on both markers and different groupings of the samples showed no evidence of sub-structuring within O-stock. A simulation exercise showed that the statistical power of the homogeneity test was high. The DAPC showed clear differentiation between J- and O-stocks but no evidence of sub-structuring within the O-stock sample [62].

A later study used DAPC and spatial analysis of principal component (sPCA) [65] to investigate population structure [66]. The study was based on msDNA (16 loci) and the sample sizes were similar to the previous study [61]. The DAPC failed to find evidence of additional structure other than the J- and O-stocks. The results indicated a low possibility that multiple stocks exist (other than the J- and O-stocks) with overlapping geographic ranges.

A different approach was used in a study that used msDNA data at 16 loci in 4554 whales to infer Parent-Offspring (P-O) relationships using a Maximum-Likelihood approach [67]. Biological information such as the sex and sexual maturity of the whales was used to interpret the genetic results on P-O pairs. The relationship between False Discovery Rate (FDR) and Power (P) was evaluated by simulation. Of 145 inferred P-O pairs (estimated FDR = 0.1), 141 were further evaluated by typing 10 additional msDNA loci. A total of 75 were confirmed (among them 26 Mother-Fetus pairs) and 66 pairs were ranked 'False Positives', yielding an overall observed FDR of 0.468. Among the validated P-O pairs, O-stock pairs were significantly overrepresented and no pairs between J- and O-stock individuals were detected. J-stock animals seem to appear on both sides of Japan closer to the coast, while O-stock individuals occur mostly to the east of Japan, both close to the coast and far offshore. The study provided no evidence for further population structure other than J and O-stocks.

Most recently, a study [68] used three spatially explicit clustering tools including GENELAND [69], TESS [70, 71] and BAPS to explore the msDNA data used previously in [66]. The authors believed that the most informative approach was GENELAND using the mixture model with correlated allele frequency model, which supported $K = 4$, i.e., four putative populations. Given the implications of this in terms of both previous analyses and management strategy evaluation, additional work was subsequently undertaken [72, 73]. That study examined the correspondence of the four above four clusters with the available genetic and non-genetic information. The authors concluded that the most plausible scenario was for two populations (J and O) with complex spatial and temporal mixing along the Pacific coast of Japan [72, 73]. They further noted that some of the analyses conducted were consistent with a scenario of coastal areas containing genetically admixed individuals, and recommended further analyses under the GENELAND as well under the TESS and BAPS.

4.4 Summary and suggestions for future work

Over the last two decades, several important genetic studies focused on investigating population identification and structure in minke whales have been undertaken in three oceanic basins using two genetic markers, mtDNA and msDNA. The driving force behind these analyses was obtaining information to help with effective conservation and management. Of necessity, all of these studies were based on genetic samples collected in feeding grounds and migratory corridors. In this context, population identification is associated with the concept of Management Units (MUs) described by one author in 1994 as 'populations with a significant divergence of allele frequencies at nuclear or mitochondrial loci, regardless of the phylogenetic distinctiveness of the alleles' [37]. Several of the studies described above presented statistical results that are consistent with this criterion for defining the population. In addition to hypothesis testing, several increasingly sophisticated clustering approaches have been used for the purpose of identifying populations.

Recent studies in the Southern Hemisphere were concentrated in the Indo-Pacific sector of the Antarctic where a large number of genetic samples of Antarctic minke whales was available from Japanese whale research programs. At least two populations have been identified in this sector, the I- and P-populations, which may be related to breeding grounds in lower latitudes of the eastern Indian Ocean and western South Pacific. These populations exhibit significant differences in their mtDNA haplotype and msDNA alleles frequencies, matching the criterion for Management Unit defined above. The Australian continent may play a role in isolating these populations during the winter breeding season, with whales presenting some degree of fidelity to particular feeding grounds in the Antarctic during summer. Although, a transition area of mixing of these two populations was postulated in the Antarctic feeding grounds, whales from each population appear to return to their respective breeding grounds in winter.

To fully understand population structure in the Southern Hemisphere, additional effort should be made to collect genetic samples from other sectors of the Antarctic and other regions of the Southern Hemisphere. This will allow investigation of the full distribution of the P- and I-populations as well the research into structure in the remaining sectors of the Antarctic. Clearly, any understanding of population structure will be greatly facilitated by dedicated efforts to investigate the migration routes and locations of breeding areas; satellite tracking will be an extremely valuable tool in this regard [74].

In the North Atlantic, the results of several genetic studies on population identification and structure may appear contradictory. While some studies suggested subtle genetic differences among groups of whales, others studies based on larger sample sizes have failed to detect any evidence of structure in this oceanic basin. As in the Southern Hemisphere, research on migratory routes and locations of breeding grounds is required to assist the interpretation of the results of the genetic analyses in the feeding ground and migratory routes.

In the North Pacific, recent genetic analyses have been concentrated in the western side due to a larger availability of genetic samples from the Japanese whale research programs and to management needs within the context of the IWC's Scientific Committee. Historically two populations have been recognized in the western North Pacific, the J- and O-stocks, and recent genetic analyses have confirmed their existence and furthermore have revealed more information on their patterns of spatial and seasonal movement. The J-stock occurs mainly in the Sea of Japan although some individuals migrate seasonally to the Pacific side of Japan. The O-stock is mainly found on the Pacific side of Japan. The objective of most recent studies has been to whether or not additional structure occurs within either or both of the J- and O-stocks, and several new analytical approaches were used to respond that question. Results of most of the approaches indicated a lack of additional structure, other than that attributed to the J- and O- stocks. The most recent IWC Scientific Committee discussions allocated high plausibility to the hypothesis of two populations with spatial/temporal mixing in the western North Pacific [75]. As for the other two ocean basins, effort should be made to collect and analyze genetic samples from the less understood eastern North Pacific as well to undertake focused research to understand migratory corridors and breeding ground locations.

It is also important to make effort to investigate the occurrence, distribution and population structure of common minke whales distributed around Chinese and Korean Peninsula waters, and the genetic relationship with whales distributed in the subareas around Japan. Investigation of the population genetic structure in those waters is important as several annual bycatches have been reported for the Korean Peninsula.

5. General conclusions

Many genetic studies on minke whales were conducted in the last 20 years. New taxonomic information post-1998 relates primarily to the Southern Hemisphere common minke whales (dwarf minke whales) from the western South Pacific and western South Atlantic, which are differentiated by both mtDNA and msDNA markers. The paraphyletic relationship between the North Atlantic and Southern Hemisphere (WSA) common minke whale has important implications for the taxonomic definition of common minke whales. Regarding population genetic structure, at least two populations of the Antarctic minke whale have been identified in the Indo-Pacific sector of the Antarctic, and at least two populations were confirmed in the western North Pacific common minke whales. In the North Atlantic genetic studies suggest that population structure, should it exist, is rather subtle. As for the North Pacific and Southern Hemisphere, analyses are hindered by a lack of knowledge (and thus samples from) breeding grounds.

The population structure of minke whales is intertwined with some degree of fidelity to specific feeding grounds. This fidelity could vary depending on changing short- and long-term environmental conditions. In the case of the Antarctic minke whales, the pattern of distribution and movement of different populations in the feeding grounds has been related with the distribution of their key prey species, the krill (*Euphausia superba*), which in turn depends on the bottom topography as well sea ice and hydrographic features [12]. A similar story has been identified for both the North Atlantic and North Pacific and it is not surprising that feeding ground distribution reflects prey distribution. Future studies on population structure and distribution of minke whales should consider information on environmental variables especially under a scenario of climate change.

Acknowledgements

We thank crew members, researchers and many other persons for the collection of genetic samples from minke whales from different sources, which made possible the studies reviewed in this chapter. We also thank Greg Donovan, former Head of Science of the International Whaling Commission, for useful comments and suggestions that improved substantially the previous version of this chapter. Our appreciation to Jorge Acevedo and Lucas Milmann for their assistance in figure drawing.

Conflict of interest

All authors declare that there is no conflict of interest.

Author details

Luis A. Pastene*, Mutsuo Goto, Mioko Taguchi and Yoshihiro Fujise
Institute of Cetacean Research, Tokyo, Japan

*Address all correspondence to: pastene@cetacean.jp

References

[1] Reeves RR, Stewart BS, Clapham PJ, Powell JA. Guide to Marine Mammals of the World. 1st ed. New York: Alfred A. Knopf Inc.; 2002. 527p

[2] Williamson GR. Three unusual rorqual whales from the Antarctic. Proceedings. Zoological Society of London. 1959;**133**:135-144

[3] van Utrecht WL, van der Spoel S. Observations on a minke whale (Mammalia, Cetacea) from the Antarctic. Zeitschrift fiir Saugetierkunde. 1962;**27**:217-221

[4] Kasuya T, Ichihara T. Some information on minke whales from the Antarctic. Scientific Reports of the Whales Research Institute Tokyo. 1965;**19**:37-43

[5] Omura H. Osteological study of the minke whale from the Antarctic. Scientific Reports of the Whales Research Institute Tokyo. 1975;**27**:1-36

[6] Best PB. External characters of southern minke whales and the existence of a diminutive form. Scientific Reports of the Whales Research Institute Tokyo. 1985;**36**:1-33

[7] Wada S, Kobayashi T, Numachi K. Genetic variability and differentiation of mitochondrial DNA in minke whales. Reports of the International Whaling Commission. 1991;**Special issue 13**:203-215

[8] Árnason U, Gullberg A, Widegren B. Cetacean mitochondrial DNA control region: Sequences of all extant baleen whales and two sperm whale species. Molecular Biology and Evolution. 1993;**70**:960-970

[9] Pastene LA, Fujise Y, Numachi K. Differentiation of mitochondrial DNA between ordinary and dwarf forms of minke whale. Reports of the International Whaling Commission. 1994;**44**:277-281

[10] Rice DW. Marine Mammals of the World. Systematics and Distribution. Society for Marine Mammalogy, Lawrence: Special Publication Number 4; 1998. 231pp

[11] Pastene LA, Goto M, Kanda N. The utility of DNA analysis for the management and conservation of large whales. Fisheries Science. 2002;**68**(SuppI):286-289

[12] Pastene LA, Goto M. Genetic characterization and population genetic structure of the Antarctic minke whale Balaenoptera bonaerensis in the Indo-Pacific region of the Southern Ocean. Fisheries Science. 2016;**82**:873-886

[13] Pastene LA, Goto M, Kanda N, Zerbini AN, Kerem D, Watanabe K, et al. Radiation and speciation of pelagic organisms during periods of global warming: The case of the common minke whale Balaenoptera acutorostrata. Molecular Ecology. 2007;**16**(7):1481-1495

[14] Pastene LA, Acevedo J, Goto M, Zerbini NA, Acuna P, Aguayo-Lobo A. Population structure and possible migratory links of common minke whales, Balaenoptera acutorostrata, in the southern hemisphere. Conservation Genetics. 2010;**11**(4):1553-1558

[15] Glover KA, Kanda N, Haug T, Pastene LA, Oien N, Goto M, et al. Migration of Antarctic minke whales to the Arctic. PLoS One. 2010;**5**(12):e15197.

[16] Glover KA, Kanda N, Haug T, Pastene LA, Oien N, Seliussen DB, et al. Hybrids between common and Antarctic minke whales are fertile and can back-cross. BMC Genetics. 2013;**14**:25-36

[17] Malde K, Seliussen BB, Quintela M, Dahle G, Besnier F, Skaug HJ, et al. Whole genome resequencing reveals diagnostic markers for investigating global migration and hybridization between minke whale species. BMC Genomics. 2017;**18**:76. DOI: 10.1186/s12864-016-3416-5

[18] Milmann L, Taguchi M, Siciliano S, Baumgarten JE, Oliveira LR, Valiati VH, et al. New genetic evidences for distinct populations of the common minke whale (Balaenoptera acutorostrata) in the southern hemisphere. Polar Biology. 2021;44:1575-1589

[19] Adachi J, Hasegawa M. MOLPHY version 2.3: Program for molecular phylogenetics based on maximum likelihood. Computer Science Monographs of the Institute of Statistical Mathematics. 1996;**28**:1-150

[20] Yang Z. PAML: A program package for phylogenetic analyses by maximum likelihood. CABIOS. 1997;**13**:555-556

[21] Strimmer K, von Haeseler A. Quartet puzzling: A quartet maximum-likelihood method for reconstructing tree topologies. Journal of Molecular Biology and Evolution. 1996;**13**: 964-969

[22] Nikaido M, Matsuno F, Hamilton H, Brownell RL Jr, Cao Y, Ding W, et al. Retroposon analysis of major cetacean lineages: The monophyly of toothed whales and the paraphyly of river dolphins. Proceedings of the National Academy of Sciences, USA. 2001;**98**:7384-7389

[23] Saitou N, Nei M. The neighbor-joining method: A new method for reconstructing phylogenetic tree. Molecular Biology and Evolution. 1987;**4**(4):406-425

[24] Cavalli-Sforza LL, Edwards AWF. Phylogenetic analysis: Models and estimation procedures. Evolution. 1967;**21**:550-570

[25] Felsenstein J. Evolutionary trees from DNA sequences: A maximum likelihood approach. Journal of Molecular Evolution. 1981;**17**:368-376

[26] Edwards AWE, Cavalli-Sforza LL. The reconstruction of evolution. Annals of Human Genetics. 1963;**27**:104-105

[27] Nielsen R, Wakeley J. Distinguishing migration from isolation: A Markov chain Monte Carlo approach. Genetics. 2001;**158**:885-896

[28] Palsbøll PJ, Bérubé M, Aguilar A, Notarbartolo di Sciara G, Nielsen R. Discerning between recurrent gene flow and recent divergence under a finite-site mutation model applied to North Atlantic and Mediterranean Sea fin whale (Balaenoptera physalus) populations. Evolution. 2004;**58**:670-675

[29] Yang Z, Rannala B. Bayesian phylogenetic inference using mtDNA sequences: A Markov chain Monte Carlo method. Molecular Biology and Evolution. 1997;**14**(7):717-724

[30] Nei M. Molecular Evolutionary Genetics. New York: Columbia University Press; 1987

[31] Weir BS, Cockerham CC. Estimating F-statistics for the analysis of population structure. Evolution. 1984;**38**:1358-1370

[32] Shriver MD, Jin L, Boerwinkle E, Deka R, Ferrel RE, Chakraborty R. A novel measure of genetic distance for highly polymorphic tandem repeat loci. Molecular Biology and Evolution. 1995;**12**:914-920

[33] Pritchard JK, Stephens M, Donnelly P. Inference of population structure using multilocus genotype data. Genetics. 2000;**155**:945-959

[34] Piry S, Alapetite A, Cornuet JM, Paetkau D, Baudouin L, Estoup A. GENECLASS2: A software for genetic assignment and first-generation migrant detection. The Journal of Heredity. 2004;**95**(6):536-539

[35] Nielsen EEG, Bach LA, Kotlicki P. HYBRIDLAB (version 1.0): A program for generating simulated hybrids from population samples. Molecular Ecology Notes. 2006;**6**:971-973

[36] Ryder OA. Species conservation and systematics: The dilemma of subspecies. Trends in Ecology & Evolution. 1986;**1**: 9-10

[37] Moritz C. Defining 'evolutionary significant units' for conservation. Trends in Ecology & Evolution. 1994;**9**:373-375

[38] Crandall KA, Bininda-Emonds ORP, Mace GM, Wayne RK. Considering evolutionary processes in conservation biology. Trends in Ecology & Evolution. 2000;**15**(7):290-295

[39] Rosel PE, Hancock-Hanser BL, Archer FI, Robertson KM, Martien KK. Examining metrics and magnitudes of molecular genetic differentiation used to delimit cetacean subspecies based on mitochondrial DNA control region sequences. Marine Mammal Science. 2017;**33**:76-100

[40] Donovan GP. A review of IWC stock boundaries. Report of the International Whaling Commission. 1991;**Special issue 13**:39-68

[41] Punt AE, Donovan GP. Developing management procedures that are robust to uncertainty: Lessons from the International Whaling Commission. ICES Journal of Marine Science. 2007;**64**(4):603-612

[42] Murase H, Palka D, Punt A, Pastene LA, Kitakado T, Matsuoka K, et al. Review of the assessment of two stocks of Antarctic minke whales (eastern Indian Ocean and western South Pacific). Journal of Cetacean Research and Management. 2020;**21**(1):95-122

[43] International Whaling Commission. Report of the intersessional workshop to review data and results from special permit research on minke whales in the Antarctic, Tokyo, 4-8 December 2006. Journal of Cetacean Research and Management (Suppl.). 2008;**10**:411-445

[44] Roff DA, Bentzen P. The statistical analysis of mtDNA polymorphisms: $\chi 2$ and the problem of small samples. Molecular Biology and Evolution. 1989;**6**:539-545

[45] Kitakado T, Schweder T, Kanda N, Pastene LA, Walloe L. Dynamic Population Segregation by Genetics and Morphometrics in Antarctic Minke Whales. Paper SC/65b/IA13 presented to the IWC Scientific Committee, Bled; 2014. 25pp. [Available from the IWC Secretariat]

[46] Ohashi Y, Goto N, Taguchi M, Pastene LA, Kitakado T. Evaluation of a paternity method based on microsatellite DNA genotypes for estimating the abundance of Antarctic minke whales (Balaenoptera bonaerensis) in the Indo-Pacific region of the Antarctic. Cetacean Population Studies. 2021;**3**:139-151

[47] Andersen LW, Born EW, Dietz R, Haug T, Oien N, Bendixen C. Genetic

population structure of minke whales *Balaenoptera acutorostrata* from Greenland, the north East Atlantic and the North Sea probably reflects different ecological regions. Marine Ecology Progress Series. 2003;**247**:263-280

[48] Anderwald P, Danielsdóttir AK, Haug T, Larsen F, Lesage V, Reid RJ, et al. Possible cryptic stock structure for minke whales in the North Atlantic: Implications for conservation and management. Biological Conservation. 2011;**144**:2479-2489

[49] Quintela M, Skaug HJ, Oien N, Haug T, Seliussen BB, Solvang HK, et al. Investigating population genetic structure in a highly mobile marine organism: the minke whale Balaenoptera acutorostrata acutorostrata in the north East Atlantic. PLoS One. 2014;**9**(9):e108640

[50] Excoffier L, Smouse PE, Quattro JM. Analysis of molecular variance inferred from metric distances among DNA haplotypes: Application to human mitochondrial DNA restriction data. Genetics. 1992;**131**:479-491

[51] Raymond M, Rousset F. GENEPOP Ver 3.2.: A population genetics software for exact tests and ecumenicism. The Journal of Heredity. 1995;**86**:248-249

[52] Dieringer D, Schlotterer C. Microsatellite analyser (MSA): A platform independent analysis tool for large microsatellite data sets. Molecular Ecology Notes. 2002;**3**(1):167-169

[53] Evanno G, Regnaut S, Goudet J. Detecting the number of clusters of individuals using the software structure: A simulation study. Molecular Ecology. 2005;**14**:2611-2620

[54] Corander J, Waldmann P, Marttinen P, Sillanpaa MJ. BAPS 2: Enhanced possibilities for the analysis of genetic population structure. Bioinformatics. 2004;**20**:2363-2369

[55] Slatkin M. A measure of population subdivision based on microsatellite allele frequencies. Genetics. 1995;**139**: 457-462

[56] Excoffier L, Laval G, Schneider S, Arlequin ver. 3.0: An integrated software package for population genetics data analysis. Evolutionary Bioinformatics Online. 2005;**1**:47-50

[57] Omura H, Sakiura H. Studies on the little piked whale from the coast of Japan. Scientific Reports of the Whales Research Institute Tokyo. 1956;**11**:1-37

[58] Kato H. Body length, reproduction and stock separation of minke whales off northern Japan. Report of the International Whaling Commission. 1992;**42**:443-453

[59] Wada S, Numachi K. Allozyme analyses of genetic differentiation among the populations and species of the Balaenoptera. Report of the International Whaling Commission 1991;**Special issue 13**:125-154

[60] Goto M, Pastene LA. Population structure of the western North Pacific minke whale based on an RFLP analysis of the mtDNA control region. Report of the International Whaling Commission. 1997;**47**:531-537

[61] Pastene LA, Goto M, Taguchi M, Kitakado T. Temporal and Spatial Distribution of the 'J' and 'O' Stocks of Common Minke Whale in Waters around Japan Based on Microsatellite DNA. Paper SC/F16/JR38 presented to the JARPNII special permit expert panel review workshop, Tokyo, February 2016. 2016;14pp. [Available from the IWC Secretariat]

[62] Pastene LA, Goto M, Taguchi M, Kitakado T. Updated Genetic Analyses Based on Mitochondrial and Microsatellite DNA Indicated no Sub-Structure of the 'O' Stock Common Minke Whale in the Western North Pacific. Paper SC/F16/JR40 presented to the JARPNII special permit expert panel review workshop, Tokyo, February 2016. 2016;19pp. [Available from the IWC Secretariat]

[63] Rousset F. Genepop'007: A complete re-implementation of the genepop software for windows and Linux. Molecular Ecology Resources. 2008;**8**:103-106

[64] Jombart T, Devillard S, Balloux F. Discriminant analysis of principal components: A new method for the analysis of genetically structured populations. BMC Genetics. 2010;**11**:94

[65] Jombart T, Devillard S, Dufour AB, Pontier D. Revealing cryptic spatial patterns in genetic variability by a new multivariate method. Heredity. 2008;**101**:92-103

[66] Taguchi M, Goto M, Pastene LA. Results of Discriminant Analysis of Principal Component (DAPC) and Spatial Analysis of Principal Component (sPCA) and Implications for the Stock Structure of Western North Pacific Common Minke Whale. Paper SC/F19/WNPM4 presented to the First Intersessional Workshop on the Implementation Review for western North Pacific minke whales, Tokyo, February 2019. 2019;16pp. [Available from the IWC Secretariat]

[67] Tiedemann R, Tiedemann MR, Goto M, Taguchi M, Pastene LA. Finding Parent-Offspring Pairs among Western North Pacific Common Minke Whales. Paper SC/67a/SDDNA01 presented to the IWC Scientific Committee, Bled, May 2017. 2017;17pp. [Available fromthe IWC Secretariat]

[68] de Jong M, Hoelzel AS. Collaborative Analysis of WNP Minke Whale Stock Structure Using the Japanese Microsatellite DNA Database and Spatially Explicit Population Structure Analyses. Paper SC/F19/WNPM2 presented to the First Intersessional Workshop on the Implementation Review for western North Pacific minke whales, Tokyo, February 2019. 2019;25pp. [Available from the IWC Secretariat]

[69] Guillot G, Mortier F, Estoup A. Geneland: A program for landscape genetics. Molecular Ecology Notes. 2005;**5**:712-715

[70] Chen C, Durand E, Forbes F, Francois O. Bayesian clustering algorithms ascertaining spatial population structure: A new computer program and a comparison study. Molecular Ecology Notes. 2007;**7**:747-756

[71] Francois O, Ancelet S, Guillot G. Bayesian clustering using hidden Markov random fields in spatial population genetics. Genetics. 2006;**174**:805-816

[72] Goto M, Taguchi M, Pastene LA. Additional Analyses to Assist the Interpretation of the Results of the Spatially Explicit Clustering Tools Used to Explore Microsatellite Data of Western North Pacific Common Minke Whale (SC/F19/WNPM02). Paper SC/F19/WNPM9 presented to the First Intersessional Workshop on the Implementation Review for western North Pacific minke whales, Tokyo, February 2019. 2019;19pp. [Available from the IWC Secretariat]

[73] Taguchi M, Goto M, Pastene LA. Genetic and Non-genetic Evidences Suggest a Low Plausibility for Western North Pacific Common Minke Whale

Stock Structure Hypothesis E. Paper SC/68a/SDDNA02 presented to the IWC Scientific Committee, Nairobi, May 2019. 2019;22pp. [Available from the IWC Secretariat]

[74] Konishi K, Isoda T, Bando T, Minamikawa S, Kleivane L. Antarctic minke whales find ice gaps along the ice edge in foraging grounds of the Indo-Pacific sector (60°E and 140°E) of the Southern Ocean. Polar Biology. 2020;**43**(4):343-357

[75] International Whaling Commission. Report of the Scientific Committee. Journal of Cetacean Research and Management (Suppl.). 2020;**13**:1-65

Chapter 3

Applications of Omics Approaches to Decipher the Impact of Contaminants in Dolphins

Reyna Cristina Collí-Dulá and Ixchel Mariel Ruiz-Hernández

Abstract

With the advent of omic technologies (genomic, transcriptomic, proteomic, metabolomic and lipidomic), it has been possible to identify global profiles of genes, proteins or metabolites in cells, tissues or organ systems at the same time. Key pathways can be identified associated with certain diseases, physiology processes or adverse effects in response to contaminants in marine organisms. This review focuses on underlining how the use of omics technology in dolphins has contributed to understanding its physiological responses and ambient stressors. They provide a basis for understanding dolphins' physiology and a means for monitoring health conditions as well as furthering ecotoxicology studies.

Keywords: omics technologies, contamination, dolphin

1. Introduction

One of the concerns in environmental matters is the continuous discharge of countless numbers of chemicals derived from human activities into aquatic systems. These include a large number of contaminants, among these commercial and industrial products (e.g., metals, industrial additives, surfactants and pesticides), personal care products, pharmaceuticals and endocrine-disrupting compounds, among others [1]. The presence of relevant concentrations in the environment has dramatic consequences to the organisms that inhabit these systems (e.g., affecting reproduction and survival), which is reflected in the decline of their populations and accumulation of pollutants [2].

A major concern about contaminants in aquatic systems is the bioaccumulation and biomagnification that can result with all organisms present in these systems including harmful effects to human health [3, 4]. Mammalian organisms, especially dolphins, are considered sentinel species for monitoring the health of coastal marine ecosystems [5, 6]. The main reason for that is (1) they are at the highest trophic level of the food chain and due to their role as predators, they can bioaccumulate contaminants, and (2) they also can live for longer periods (more than 40 years). It makes them good organisms to show long-term accumulation characteristics from contaminants like heavy metals in the marine environment [7]. Recently, with the development of new technologies within the "omic sciences" such as genomics, transcriptomics, proteomics and metabolomics, great advances have been made in the biological science disciplines, particularly in human health. In environmental areas "omics" have begun to have a large impact [8], mainly in aquatic toxicology [9, 10]. Together, new genomic sequencing and postgenomic technologies make it possible to obtain detailed information on drugs, toxicants, pollutants, nutrients and physical and psychological stressors on an omic scale [11]. The use of these omic technologies has allowed the emergence of ecotoxicogenomic disciplines [12, 13].

With these technologies, it is possible to determine the effect of a particular event in the life of a cell, organ or organism in response to contaminants. Through the characterization of the transcriptome, proteome or metabolome, one can perform global analysis to determinate transcriptional/proteomic or metabolomic changes at the same time in many samples (cells, tissues, biofluids, etc.) and be able to make the comparison among them. Omics technologies in environmental matters can help to assess the health statuses of aquatic systems, understand the mechanisms of action of the contaminants, through profiling of genes, proteins or metabolites that may enrich key pathways (molecular or biochemical). It sheds light on how dolphins respond to contaminants while helping to predict adverse effects on other marine organisms (**Figure 1**). This review highlights the

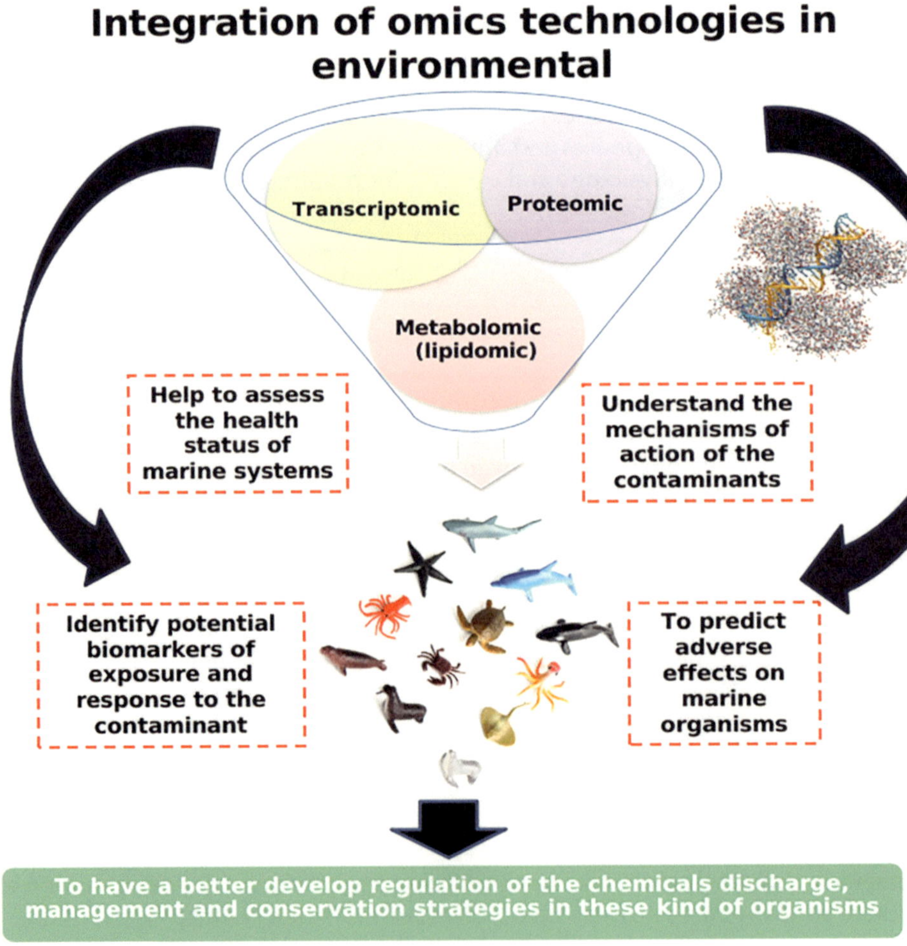

Figure 1.
Integration of omics technologies in marine organisms.

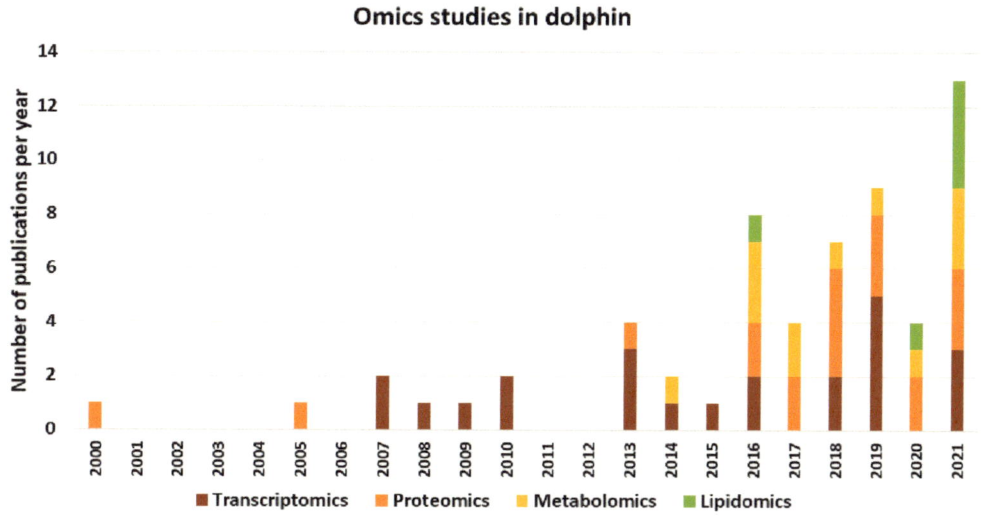

Figure 2.
Number of studies related with omics approach in dolphin.

omics studies performed on dolphins to gather information regarding contamination levels and their effects on worldwide dolphin populations (**Figure 2**). Applications of the omics approach help to understand the dolphins' physiology as a way to monitor dolphin health conditions and to further ecotoxicology studies. Conclusively, it might provide a method for developing regulations for chemical discharge as well as management and conservation strategies for these kinds of ecosystems.

The three major omics technologies that have proven to have a tremendous impact include transcriptomics, proteomics and metabolomics [14].

2. Selection of bibliographic material

We reviewed relevantly and recently published studies on the applicability and usefulness of Omics in dolphins. The selection of scientific publications was made through the use of search engines from Google Scholar, PubMed and Scopus to locate studies of interest using the keywords "transcriptomic", "proteomic", "metabolomic", "lipidomic" and "dolphin". We excluded all repeated studies. It was inevitable that some omics research was bot captured due to them not included the keywords we used. For the selection of publications, only the research studies that were related to contaminants were included in the data set (**Table 1**).

Based on these criteria, 59 publications were selected from >250 reviewed. Transcriptomics was the most frequently applied technique (38%) followed by proteomics (30%), metabolomics (21%) and finally, lipidomic (10.2%). In general searching the omic studies selected, we identified 8 topics including "contamination", "physiology" and "health" among others based on the type of research described in the publications. More details about each topic are given in **Figure 3**. Contamination studies were dominant using the transcriptomic method (31%), compared to studies focusing on proteomics (0%), metabolomics (8%) and lipidomics (0%). We noticed that proteomics and lipidomics are less used in studies related to contamination. However, proteomics is the most frequent technology applied to identifying responses associated with the physiology of dolphins (28%), followed by lipidomics (33%),

Specie	Omic approach	Type of sample	Contaminant/ stressor	Genes/Proteins/Metabolites/Lipids	Contribution	Reference
Transcriptomic						
Tursiops truncatus	Microarray (Custom 4x44K Agilent oligo array)	Cultures Skin	*Ex vivo* assay of BPA (0.1 or 1 μg/ml) and PFOA (0.1 or 1 μg/ml)	*BPA:* Genes involved response to stress (e.g., programmed cell death protein, tumor suppressor), immune system (e.g., complement factor H, class I histocompatibility alpha chain), lipid metabolism (e.g., adipogenesis regulatory factor, fatty aldehyde dehydrogenase isoform 2), and embryonic development and growth (e.g., titin, nuclear distribution protein nude-like 1). *PFOA:* Genes involved response to stress (uv excision repair protein rad23 homolog a, heat shock protein 90), immune system (complement c5, acidic mammalian chitinase), embryonic development (vascular endothelial growth factor, Rho GTPase-activating protein)	*Pathways affected: BPA:* Fat (blubber) differentiation	[15]
T. truncatus	Microarray (Custom 4X44K Agilent oligo array)	Blood	55 PCBs congeners	Genes involved in cell cycle checkpoint and apoptosis, DNA damage and chromatin remodeling (e.g., DDB1, DCN and INO80). Pathway of cellular response to stress	PCBs could cause epigenetic response, DNA damage and chromatin remodeling	[16]
T. truncatus	Microarray (Custom 4X44K Agilent oligo array)	Blood from male and female	PCBs	*Male:* Genes involved in (1) development & differentiation (e.g., SOS1), signaling pathway to maintain cell growthand survival in thyroid cells (e.g., RAS genes), (2) Wound healing & anti-tumorigenic (e.g., DCN), (3) Inflammatory Response (IL23), (4) Xenobiotic metabolism (OXR1). *Female:* Genes involved in (1) Transcription/ Translation (Maf1 homolog), (2) Immune response (e.g., tyrosine–protein kinase JACK1, 3) Development/cell growth (PAR), BB1, (TRIP11), 4) Xenobiotic metabolism (OXR1).	The development and application of a microarray to monitor global gene expression in dolphin in response to contaminants	[17]

DOI: http://dx.doi.org/10.5772/intechopen.102424

Specie	Omic approach	Type of sample	Contaminant/ stressor	Genes/Proteins/Metabolites/Lipids	Contribution	Reference
T. truncatus	RNA-seq	Skin from two ecotypes of dolphin: offshore and coastal	HOC	Genes: AHR, CYPIB1, IL16, ESR2, ESRRA, THRA. GO terms: xenobiotic metabolism, immune response, hormone metabolism, DNA repair, and metal binding.	It provides novel insight into contaminant exposure in two bottlenose dolphin ecotypes in the Southern California Bight and highlights potential relationships between HOC exposure and molecular biomarkers	[18]
T. truncatus	RNA-seq	Peripheral blood mononuclear cell (PBMC) from dolphin and human	PFOA y PFOS	*In both species:* Overexpression of genes linked to inflammation and autoimmune diseases. *Difference between species:* In human the interferon Signaling pathway is negatively regulated while in dolphin it is positively regulated. Dolphins lack Mx1 and Mx2, key proteins of the Interferon signaling pathway	This study provide a better understanding of the adverse effects of CECs (PFOA, PFOS) on both dolphin and human species	[19]
Metabolomic and lipidomic						
T. truncatus	LC/MS	Exhaled breathe	Oil spill	Phosphatidic acid, phosphatidylethanolamine, and steroids (higher abundance or uniquely in dolphins of contaminated area). Phosphatidylglycerol.	Pathways affected: • Cellular bilayer degradation • DNA and cellular damage processes • Immunological protection	[20]

Table 1.
Depict of the most relevant transcriptomic, proteomic and metabolomic studies performed in dolphin in response to the contaminants.

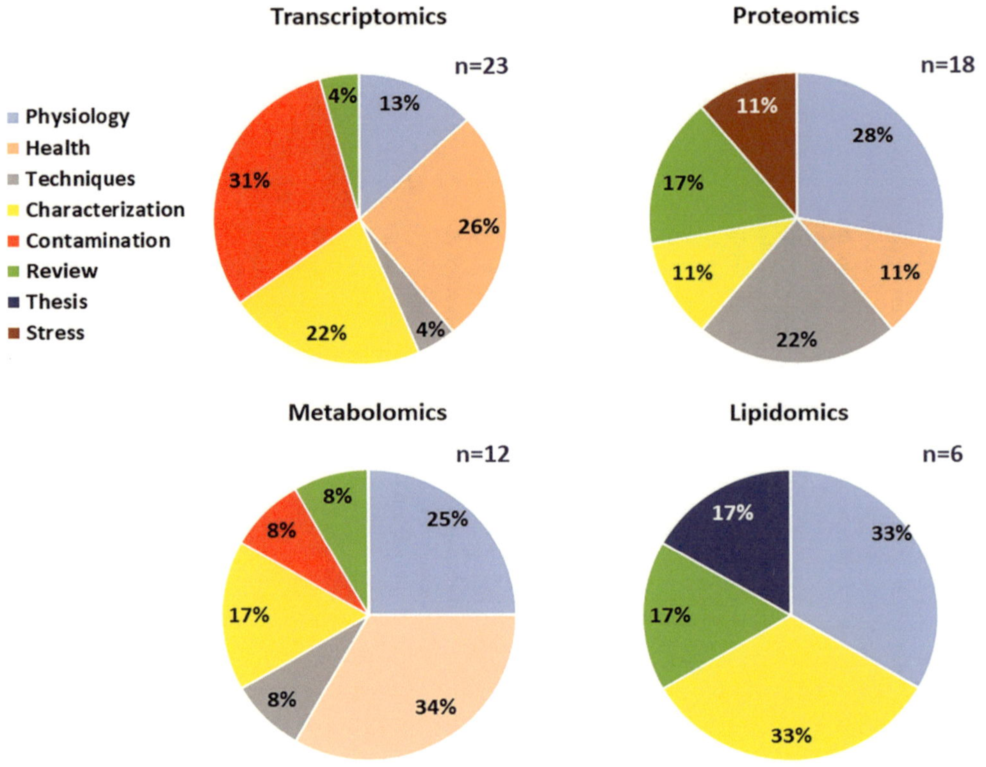

Figure 3.
Status of studies of omics in dolphins.

metabolomics (25%), and transcriptomics (13%). With respect to studies related to health, metabolomic tools (34%) were predominant, followed by transcriptomics (26%) and proteomics (11%). Interestingly, we noticed that the number of studies selected in omics and dolphins does not show an increase over time as we expected, it was diverse (**Figure 1**). After 2016, the selected literature showed an increase in the application of omics in dolphin research, notably, most studies focused on using metabolomics (LC/MS) and transcriptomic high throughput RNA sequencing (RNA-seq) tools as a diagnostic method for the detection of contaminants in oil spills and with contaminants of emerging concern (CECs). In general, it seems that there is a trend toward the increased use of transcriptomics, with studies dominating the literature from 2018 to 2019, and lipidomic applications from 2020 to 2021.

3. Omics technologies in marine organism: response to contaminants in dolphins

Omic approaches bring an integrated view of the molecules that compose a cell, tissue, or organisms in any target biological sample from a model or non-model organism. Notably, there is little information focused on proteomics, metabolomics, and lipidomics to investigate the impact of contaminants in dolphins species (**Figure 2**). We present a summary of the application of the three main omic technologies in dolphins associated with contaminants.

3.1 Transcriptomics

Transcriptomics has been the omic technique most used in biological areas because it represents all RNA molecules (e.g., miRNA, snoRNA), including the messenger RNA (mRNA) which constitutes the building blocks for translating DNA into amino acids to form proteins. The totality of mRNA is a reflex of the genes that are actively expressed in a cell or an organism at a given time and during a specific event. It permits deciphering how organisms respond to changes in the external environment or the presence of the contaminants [21]. The principal gene expression profiling methods used in transcriptomic are microarray and RNA-sequencing (RNA-seq). The difference between the potential of each method becomes apparent once the target sequences go beyond known genomic sequences. Hybridization-based techniques like microarray rely on and are limited to the transcripts bound to the array slides. Limitations of microarrays are due to the bioinformatic data available for the model organism's genome and transcriptome. RNA-seq can detect annotated transcripts but also novel sequences and splice variants [22]. RNA-seq is considered a revolutionary tool for transcriptomics in non-model organisms and is powerful enough to explore the mammalian transcriptome which was not possible with microarrays [23].

With regard to the transcriptomic studies in dolphins and contaminants, there are few studies that have used microarray methods to identify genes and molecular pathways altered by bisphenol A [2,2 bis(4-hydroxyphenyl) propane (BPA), perfluoroalkyl substances (PFAS) and perfluorooctanoic acid (PFOA) in dolphin skin biopsies [15]. These contaminants can cause changes in key genes involved in pathways related to stress, immune response, development and lipid metabolism. Likewise, there are another two studies that describe the construction and validation of the use of microarrays in *T. truncatus* as well as using bioinformatic tools to detect polychlorinated biphenyls (PCBs) from dolphin blood during the monitoring of high-level contamination at Superfund sites on the Georgia coast in the US [16, 17], see **Table 1**. A limited range of sequencing data is available for dolphins from whole-genome assemblies to RNA-seq data [18, 19], however, two studies have documented the effects of halogenated organic contaminants (HOCs) at transcriptomic levels. For example, Trego and colleagues reported that 20 skin biopsys from *T. truncatus* dolphin collected on the Southern California Bight showed to have a positive correlation with the presence of HOCs and genes associated with the metabolism of xenobiotics and with the immune and endocrine pathways. Likewise, in another study also performed with *T. truncatus*, human peripheral blood mononuclear cells (PBMC) from both species were assessed to investigate the effects of contaminant exposures of CECs (PFAs; PFOA and perfluorooctane sulfonate (PFOS)) using RNA-seq. Transcriptomic analysis showed that in both human and dolphin pathways related with endocrine immune system that inflammatory responses increased (**Table 1**) [19].

3.2 Proteomics

The main focus of proteomics is to identify and quantify all protein content in a cell, tissue, or organism and understand their functions, structure and their modifications in response to external stimuli [24]. Based on proteomics, baseline studies have been conducted to characterize proteins from spermatozoa and seminal plasma in bottlenose dolphins [25] which has been used in zooarchaeology for species identification of cetaceans [26]. Other studies have been focused on developing bioinformatics tools or methods to obtain or analyze proteins from different samples [27–29].

Most of the proteomic studies in dolphins have focused on the physiology of proteins and peptides. These studies have provided valuable information, such as the case of the proline-rich antimicrobial peptides found in different cetacean species, where these peptides could provide useful insights for future antibiotics [30]. Through proteomics one can also identify peptides related to metabolic disorders [31] and biomarkers of infection for diagnosis of aspergillosis in dolphins [32]. Thanks to proteomics, it has been possible to identify stress proteins involved in apoptosis, proteotoxicity and inflammation on managed and wild dolphins and their relation with biological data such as serological, biochemical, hematological and endocrine variables [33]. In stressed cetaceans, 30 stress-activated proteins have been identified, where these proteins have an important role in cellular detoxification, stress response, cell growth and differentiation, apoptosis, immunologic, neurologic and hormonal signaling and oxidative stress response [34].

In toxicology, proteomic studies are important because the proteome is the link between effects at the molecular and the whole organism level and provide snapshot functional information of a cell under certain conditions, and it allows the identification of new biomarkers and pathways of toxicity [35]. However, studies related to contamination have not been reported yet.

Regarding the methods and tools used in proteomics, initially, the way to analyze variations of protein expression was by gel electrophoresis. Now the main tool used is mass spectrometry with their different techniques: LC/MS, MALDI TOF/TOF, ESI-QUAD-TOF, iTRAQ. Protein microarray has also been used for these kinds of studies and bioinformatic tools.

Proteomics generates a large amount of data that permit furthering one's knowledge of mechanisms of action and toxicant effect of a contaminant in organisms and thus be able to understand biological processes [35]. However, the limitations in these kinds of studies are with peptide separations, identification and that many species lack of protein sequence information [14, 36].

3.3 Metabolomics

Metabolomics is responsible for identifying and quantifying all endogenous and exogen metabolites in an organism or biological sample [37]. Metabolites are all final products of cellular processes and knowing their levels permits one to understand the responses of a biological system to environmental changes [38].

This omic tool contributes to understanding of how environmental stressors can affect human and environmental health. However, these kinds of applications have not yet been explored as often in dolphins [39]. Most of the metabolomic studies in dolphins have been focused on establishing baseline information on health [40–43], and physiology [44–46] with a few studies looking at the characterization of metabolites from exhaled breath and tears [47, 48].

Regarding pollution studies, just only a single work was discovered. After the spill of the Deepwater Horizon in the Gulf of Mexico, dolphin populations were severely affected, showing adrenal and lung diseases, poor reproductive success and higher mortality [49–51]. In bottlenose dolphins, *Tursiops truncatus* exhaled breathe metabolites had been studied [20] from a managed collection in San Diego, from a wild population in Sarasota Bay and Barataria Bay, the latter being the contaminated site. Several metabolites, such as yiamoloside B, diacylglycerol, leptomycin B, phosphatidylglycerol and phospholipids, were correlated with pulmonary disease. Cortisol and aldosterone levels were lower in Barataria Bay, also dolphins from this population presented thin adrenal gland cortices, supporting an impaired hypothalamus-pituitary-adrenal axis. Lower amounts of glucose in the contaminated area may represent a response to stress or feeding. Besides, metabolites as steroids, phosphatidic acid and phosphatidylethanolamine were unique or found in higher abundance in the contaminated area compared to the healthy reference dolphins which suggest cellular destruction. Many of the specific metabolites found in dolphins from Barataria Bay, were markers for arachidonic acid, lipid oxidation and lung surfactant breakdown. In addition, antibiotics, such as jadomycin B, leukomycin A1 and A7, lansonolide A, chivosazole E and mycolacton, were also found in dolphins from Barataria Bay. These compounds are products of fungi and bacteria suggesting that dolphins exposed to oil spill may have pneumonia.

In metabolomics, the main tools used for analysis are mass spectrometry with their different instrumentation: chromatography/mass spectrometry (GC/MS), liquid chromatography/mass spectrometry (LC/MS) and in tandem (LC/MS/MS), high-performance liquid chromatography (HPLC), HPLC-MS/MS, reverse phase chromatography (RP)/UPLC-MS/MS, capillary electrophoresis time of flight mass spectrometer (CE-TOFMS), liquid chromatography/time-of-flight/mass spectrometry (LC-TOFMS) and the least used are nuclear resonance magnetic (NMR) and high-resolution magic angle spinning (HR-MAS) NMR spectroscopy.

Metabolomics is relatively a new tool and captures more integrated information of the physiology of an organism than transcriptomics or proteomics [52] because it represents the final cellular signaling events, resulting from transcriptional and translational changes [39]. However, it presents some limitations such as targeting metabolites that are species specific as well as libraries and software programs that are not yet sufficiently extensive [52].

3.3.1 Lipidomics

Lipidomics is a specialized subfield of metabolomics. Through lipidomics, it is possible to characterize all lipids from a cell, tissue, fluid, etc. and understand how these lipids influence a biological system and participate in several processes as well as how they interact with other molecules and respond to environmental changes [53, 54]. Lipids represent a major component of the metabolome [54], have an important role as components of cell membranes and participate in many cellular pathways and due to these being involved in many physiological mechanisms, also are excellent candidates for monitoring the effects of stress [55].

One representative area in marine mammals is their blubber. This is the most important site of fat and energy storage and also participates in different processes such as insulation, thermoregulation and buoyancy and, it represents up to 50% of the body mass [56] and due to the great quantity of lipids, it makes it a good repository for contaminants that are lipophilic [57]. For these reasons, lipidomics makes an excellent tool for studying the effects of contamination in these sentinel species. Although lipidomic studies have been increasing in recent years, until now, there are no dolphin lipidomic studies related to contamination. Indirectly, one study focused on respiratory metabolites [20], where some lipids were detected, including phosphatidylethanolamine, from oil spill exposure. These lipids were found in higher concentrations in dolphins from the contaminated area.

Few lipidomic studies have been reported, with most focused on physiology [58, 59], and characterization of lipids from cardiac phospholipidome [59] of small cetaceans and lipids from the blubber of killer whales [60].

The main tool used for lipidomic studies is mass spectrometry. This analysis generally uses another instrument such as LC-electrospray ionization (ESI) quadrupole time-of-flight (Q-TOF), liquid chromatography-high-resolution mass spectrometry (LC/HRMS/MS), GC-MS and LC-MS/MS and hydrophilic interaction liquid chromatography-mass spectrometry (HILIC-LC-MS).

4. Omics and dolphin: future considerations

This review integrates the available information on the effects of pollution using an omics approach on dolphins and other cetaceans considered as ideal organisms to assess and monitor pollution in coastal or ocean systems. Although there are wide applications of omic approaches in other model and non-model aquatic organisms involving environmental matters, there are very few studies from an omic perspective in dolphins. There is much evidence in the literature of the analytical power that these tools have their contribution in providing relevant information on the MOA of contaminants in cells, tissues, organisms or populations to help to assess the health status of marine systems, to identify potential biomarkers of exposure and response to the contaminant as well to predict adverse effects on marine organisms. Information provided from this study may be useful for risk assesment analysis that may impact future environmental regulations. However, there are still several limitations that need addressing in their application in dolphins. (1) One of the main challenges is with sampling (non-invasive/biopsy). There are prohibitive costs and time delays associated with obtaining the permits required to obtain samples in wildlife organisms in some countries. A non-ideal but possible option is the sampling of strandings. (2) The application of omic studies in ecotoxicology still has many challenges. The increase of these studies at different omic levels has grown impressively thus requiring improved bioinformatics and computational tools for better analysis regarding environmental stressors, such as pollutants. (3) Likewise, the collaboration between academic government entities and industry still needs to be improved.

5. Conclusions

This review highlights the importance of omic studies in dolphins which have contributed greatly in recognizing the presence and effect of contaminants such as HOC, CECs (BPA and PFOs) and those associated with oil spills (summarized in **Table 1**). Omics technologies are important to study adverse effects of contaminants or environmental changes because they provide information on the alterations of genes, proteins, metabolites and phenotypic responses [14]. Transcriptomic-based investigations were used most frequently (31%); only a few studies used a metabolomic approach (8%). The principal tool used for transcriptomic is RNA-seq and for proteomics, metabolomics and lipidomics is mass spectrometry coupled to different types of spectrometers (**Figure 1**).

Some of the more likely applications for omics in dolphins are characterization and physiology. Although omics studies have been used for many topics, the number of studies concerning contamination is rather low. Studies of proteomics, metabolomic and lipidomic are still lacking; therefore, these findings may give insight for future studies. This type of study contributes greatly in establishing baselines for environmental health studies of coastal and marine systems, the health status of the dolphin reflects the status of their environment. Perhaps it may allow the local as well as the scientific community to be more aware of marine ecosystem conditions and to recognize the importance and possibilities of integrate omics studies regarding pollution.

Conflict of interest

The authors declare no conflict of interest.

Author details

Reyna Cristina Collí-Dulá[1,2]* and Ixchel Mariel Ruiz-Hernández[2]

1 CONACYT, Ciudad de México, México

2 Departamento de Recursos del Mar, Centro de Investigación y de Estudios Avanzados del IPN, Unidad Mérida, México

*Address all correspondence to: rcolli.dula@cinvestav.mx

References

[1] Song L, Lam PKS, Hecker M. Aquatic toxicology: History and future. Aquatic Toxicology. 2019;**216**(105326):1-2

[2] Connell DW. Bioaccumulation of Xenobiotic Compounds. Boca Ratón, Florida: CRC Press; 1938

[3] Zenker A, Cicero MR, Prestinaci F, Bottoni P, Carere M. Bioaccumulation and biomagnification potential of pharmaceuticals with a focus to the aquatic environment. Journal of Environmental Management. 2014;**133**:378-387. DOI: 10.1016/j.jenvman.2013.12.017

[4] Ali H, Khan E. Trophic transfer, bioaccumulation, and biomagnification of non-essential hazardous heavy metals and metalloids in food chains/webs—Concepts and implications for wildlife and human health. Human and Ecological Risk Assessment. 2019;**25**(6):1353-1376. DOI: 10.1080/10807039.2018.1469398

[5] Bossart GD. Marine mammals as sentinel species for oceans and human health. Veterinary Pathology. 2011;**48**(3):676-690

[6] Wells RS, Rhinehart HL, Hansen LJ, Sweeney JC, Townsend FI, Stone R, et al. Bottlenose dolphins as marine ecosystem sentinels : Developing a health monitoring system. EcoHealth. 2004;**1**:246-254

[7] Honda K, Tatsukawa R. Distribution of cadmium and zinc in tissues and organs, and their age-related changes in striped dolphins, *Stenella coeruleoalba*. Archives of Environmental Contamination and Toxicology. 1983;**12**:503-512

[8] Ge Y, Wang DZ, Chiu JF, Cristobal S, Sheehan D, Silvestre F, et al. Environmental OMICS: Current status and future directions. Journal of Integrated OMICS. 2013;**3**(2):75-87

[9] Zhang X, Xia P, Wang P, Yang J, Baird DJ. Omics advances in ecotoxicology. Environmental Science & Technology. 2018;**52**:3842-3851

[10] Simmons DBD, Benskin JP, Cosgrove JR, Duncker BP, Ekman DR, Martyniuk CJ, et al. Omics for aquatic ecotoxicology: Control of extraneous variability to enhance the analysis of environmental effects. Environmental Toxicology and Chemistry. 2015;**34**(8):1693-1704

[11] Niedzwiecki MM, Walker DI, Vermeulen R, Chadeau-Hyam M, Jones DP, Miller GW. The exposome: Molecules to populations. Annual Review of Pharmacology and Toxicology. 2019;**59**:107-127

[12] Snape JR, Maund SJ, Pickford DB, Hutchinson TH. Ecotoxicogenomics: The challenge of integrating genomics into aquatic and terrestrial ecotoxicology. Aquatic Toxicology. 2004;**67**(2):143-154

[13] Poynton HC, Wintz H, Vulpe CD. Progress in ecotoxicogenomics for environmental monitoring, mode of action, and toxicant identification. Advances in Experimental Biology. 2008;**2**:21-73

[14] Amiard-Triquet C, Amiard J-C, Mouneyrac C. Aquatic ecotoxicology. In: Advancing Tools for Dealing with Emerging Risks. New York: Academic Press; 2015

[15] Lunardi D, Abelli L, Panti C, Marsili L, Fossi MC, Mancia A. Transcriptomic analysis of bottlenose

dolphin (*Tursiops truncatus*) skin biopsies to assess the effects of emerging contaminants. Marine Environmental Research. 2016;**114**(Jan):74-79. DOI: 10.1016/j.marenvres.2016.01.002

[16] Mancia A, Ryan JC, Van Dolah FM, Kucklick JR, Rowles TK, Wells RS, et al. Machine learning approaches to investigate the impact of PCBs on the transcriptome of the common bottlenose dolphin (*Tursiops truncatus*). Marine Environmental Research. 2014;**100**:57-67. DOI: 10.1016/j.marenvres.2014.03.007

[17] Mancia A, Abelli L, Kucklick JR, Rowles TK, Wells RS, Balmer BC, et al. Microarray applications to understand the impact of exposure to environmental contaminants in wild dolphins (*Tursiops truncatus*). Marine Genomics. 2015;**19**:47-57. DOI: 10.1016/j.margen.2014.11.002

[18] Trego ML, Hoh E, Whitehead A, Kellar NM, Lauf M, Datuin DO, et al. Contaminant exposure linked to cellular and endocrine biomarkers in Southern California bottlenose dolphins. Environmental Science & Technology. 2019;**53**:3811-3822

[19] Wolf BJ, Kamen DL, Fair P, Hardiman G. Meta-analysis of dolphin and human peripherial blood mononuclear cells reveals inflammatory signatures associated with exposure to high levels of perfluoroalkyl substances. International Journal of Advances in Science, Engineering and Technology. 2021;**7**(3):66-72

[20] Pasamontes A, Aksenov AA, Schivo M, Rowles T, Smith CR, Schwacke LH, et al. Noninvasive respiratory metabolite analysis associated with clinical disease in cetaceans: A deepwater horizon oil spill study. Environmental Science & Technology. 2017;**51**(10):5737-5746

[21] Schirmer K, Fischer BB, Madureira DJ, Pillai S. Transcriptomics in ecotoxicology. Analytical and Bioanalytical Chemistry. 2010;**397**:917-923

[22] Mantione KJ, Kream RM, Kuzelova H, Ptacek R, Raboch J, Samuel JM, et al. Comparing bioinformatic gene expression profiling methods: Microarray and RNA-Seq. Medical Science Monitor Basic Research. 2014;**20**:138-142

[23] Wang Z, Gerstein M, Snyder M. RNA-Seq: A revolutionary tool for transcriptomics. Nature Reviews Genetics. 2009;**10**:57-63

[24] Aslam B, Basit M, Nisar MA, Khurshid M, Rasool MH. Proteomics: Technologies and their applications. Journal of Chromatographic Science. 2017;**55**(2):182-196

[25] Fuentes-Albero MC, González-Brusi L, Cots P, Luongo C, Abril-Sánchez S, Ros-Santaella JL, et al. Protein identification of spermatozoa and seminal plasma in bottlenose dolphin (*Tursiops truncatus*). Frontiers in Cell and Development Biology. 2021;**9**:1-17

[26] Biard V, Gol'din P, Gladilina E, Vishnyakova K, McGrath K, Vieira FG, et al. Genomic and proteomic identification of Late Holocene remains: Setting baselines for Black Sea odontocetes. Journal of Archaeological Science: Reports. 2017;**15**:262-271

[27] Yu J, Kindy MS, Ellis BC, Baatz JE, Peden-Adams M, Ellingham TJ, et al. Establishment of epidermal cell lines derived from the skin of the Atlantic bottlenose dolphin (*Tursiops truncatus*). The Anatomical Record. Part A, Discoveries in Molecular, Cellular, and Evolutionary Biology. 2005;**287A**:1246-1255

[28] Dooley CT, Ferrer T, Pagán H, O'Corry-Crowe GM. Bridging immunogenetics and immuno proteomics: Model positional scanning library analysis for major histocompatibility complex class II DQ in *Tursiops truncatus*. PLoS One. 2018;**13**(8):e0201299

[29] Bergfelt DR, Lippolis J, Vandenplas M, Davis S, Miller BA, Madan R, et al. Preliminary analysis of the proteome of exhaled breath condensate in bottlenose dolphins (*Tursiops truncatus*). Aquatic Mammals. 2018;**44**(3):256-266

[30] Sola R, Mardirossian M, Beckert B, De Luna LS, Prickett D, Tossi A, et al. Characterization of cetacean proline-rich antimicrobial peptides displaying activity against eskape pathogens. International Journal of Molecular Sciences. 2020;**21**:1-17

[31] Neely BA, Carlin KP, Arthur JM, McFee WE, Janech MG. Ratiometric measurements of adiponectin by mass spectrometry in bottlenose dolphins (*Tursiops truncatus*) with iron overload reveal an association with insulin resistance and glucagon. Frontiers in Endocrinology (Lausanne). 2013;**4**(132):1-13

[32] Desoubeaux G, Piqueras MDC, Le-Bert C, Fravel V, Clauss T, Delaune AJ, et al. Labeled quantitative mass spectrometry to study the host response during aspergillosis in the common bottlenose dolphin (*Tursiops truncatus*). Veterinary Microbiology. 2019;**232**:42-49

[33] Wilson AE, Fair PA, Carlson RI, Houde M, Cattet M, Bossart GD, et al. Environment, endocrinology, and biochemistry influence expression of stress proteins in bottlenose dolphins. Comparative Biochemistry and Physiology. Part D, Genomics & Proteomics. 2019;**32**:100613

[34] Southern S. Molecular analysis of stress-activated proteins and genes in dolphins and whales: A new technique for monitoring environmental stress. In: Annual Conference-American Association of Zoo Veterinarians. Vol. 2000. Houston, TX: American Association of Zoo Veterinarians; 1998. pp. 240-244

[35] Lemos MFL, Soares AMVM, Correia AC, Esteves AC. Proteins in ecotoxicology—How, why and why not? Proteomics. 2010;**10**:873-887

[36] Monsinjon T, Knigge T. Proteomic applications in ecotoxicology. Proteomics. 2007;**7**:2997-3009

[37] Idle JR, Gonzalez FJ. Metabolomics. Cell Metabolism. 2007;**6**:348-351

[38] Fiehn O. Metabolomics—The link between genotypes and phenotypes. Plant Molecular Biology. 2002;**48**:155-171

[39] Miller MG. Environmental metabolomics: A SWOT analysis (strengths, weaknesses, opportunities, and threats). Journal of Proteome Research. 2007;**6**:540-545

[40] Venn-Watson S, Baird M, Novick B, Parry C, Jensen ED. Modified fish diet shifted serum metabolome and alleviated chronic anemia in bottlenose dolphins (*Tursiops truncatus*): Potential role of odd-chain saturated fatty acids. PLoS One. 2020;**15**(4):1-23

[41] Ardente A, Garrett T, Wells R, Walsh M, Smith C, Colee J, et al. A targeted metabolomics assay to measure eight purines in the diet of common bottlenose dolphins, *Tursiops truncates*. Journal of Chromatography and Separation Techniques. 2016;**7**(5):1-22

[42] Borras E, Aksenov AA, Baird M, Novick B, Schivo M, Zamuruyev KO, et al. Exhaled breath condensate methods adapted from human studies using longitudinal metabolomics for predicting early health alterations in dolphins. Analytical and Bioanalytical Chemistry. 2017;**409**:6523-6536

[43] Houser DS, Derous D, Douglas A, Lusseau D. Metabolic response of dolphins to short-term fasting reveals physiological changes that differ from the traditional fasting model. The Journal of Experimental Biology. 2021;**224**:1-12

[44] Suzuki M, Yoshioka M, Ohno Y, Akune Y. Plasma metabolomic analysis in mature female common bottlenose dolphins: Profiling the characteristics of metabolites after overnight fasting by comparison with data in beagle dogs. Scientific Reports. 2018;**8**:1-11

[45] Misra BB, Ruiz-Hernández IM, Hernández-Bolio GI, Hernández-Núñez E, Díaz-Gamboa R, Colli-Dula RC. 1H NMR metabolomic analysis of skin and blubber of bottlenose dolphins reveal a functional metabolic dichotomy. Comparative Biochemistry and Physiology Part D: Genomics and Proteomics. 2019;**30**:25-32. DOI: 10.1016/j.cbd.2019.02.004

[46] Velasco-Martínez I del C, Hernández-Camacho CJ, Méndez-Rodríguez LC, Zenteno-Savín T. Purine metabolism in response to hypoxic conditions associated with breath-hold diving and exercise in erythrocytes and plasma from bottlenose dolphins (*Tursiops truncatus*). Comparative Biochemistry and Physiology Part A: Molecular & Integrative Physiology. 2016;**191**:196-201

[47] Aksenov AA, Yeates L, Pasamontes A, Siebe C, Zrodnikov Y, Simmons J, et al. Metabolite content profiling of bottlenose dolphin exhaled breath. Analytical Chemistry. 2014;**86**:10616-10624

[48] Pinazo-DuranMD,Sanz-GonzalezSM, Valero-Vello M, Garcia-Medina JJ, Benitez del Castilo JM. Characterization of dolphin tear metabolomics. Investigative Ophthalmology & Visual Science. 2021;**62**(8):1557

[49] Venn-Watson S, Colegrove KM, Litz J, Kinsel M, Terio K, Saliki J, et al. Adrenal gland and lung lesions in Gulf of Mexico common bottlenose dolphins (*Tursiops truncatus*) found dead following the Deepwater horizon oil spill. PLoS One. 2015;**10**(5):1-23

[50] Lane SM, Smith CR, Mitchell J, Balmer BC, Barry KP, McDonald T, et al. Reproductive outcome and survival of common bottlenose dolphins sampled in Barataria bay, Louisiana, USA, following the Deepwater horizon oil spill. Proceedings of the Royal Society B: Biological Sciences. 2015;**282**:1-9

[51] Venn-Watson S, Garrison L, Litz J, Fougeres E, Mase B, Rappucci G, et al. Demographic clusters identified within the northern Gulf of Mexico common bottlenose dolphin (*Tursiops truncates*) unusual mortality event: January 2010–June 2013. PLoS One. 2015;**10**(2):1-13

[52] Ankley GT, Daston GP, Degitz SJ, Denslow ND, Hoke RA, Kennedy SW, et al. Toxicogenomics in regulatory ecotoxicology. Environmental Science & Technology. 2006:4055-4065

[53] Watson AD. Lipidomics: A global approach to lipid analysis in biological systems. Journal of Lipid Research. 2006;**47**:2101-2111

[54] Wood P. In: Wood P, editor. Lipidomics. New York, NY: Humana Press; 2017. p. 248

[55] Aristizabal-Henao JJ, Ahmadireskety A, Griffin EK, Ferreira Da Silva B, Bowden JA. Lipidomics and environmental toxicology: Recent trends. Current Opinion in Environmental Science & Health. 2020;**15**:26-31. DOI: 10.1016/j.coesh.2020.04.004

[56] Iverson SJ. Blubber. In: Perrin WF, Würsig B, JGM T, editors. Encyclopedia of Marine Mammals. 2nd ed. Academic Press; 2009. pp. 115-120. DOI: 10.1016/B978-0-12-383832-2.00112-2

[57] Reddy ML, Dierauf LA, Gulland FMD. Marine mammals as sentinels of ocean health. In: Dierauf LA, Gulland FMD, editors. Handbook of Marine Mammal Medicine. 2nd ed. Boca Ratón, FL: CRC Press; 2001. pp. 3-13

[58] Sobolesky PM, Harrell TS, Parry C, Venn-Watson S, Janech MG. Feeding a modified fish diet to bottlenose dolphins leads to an increase in serum adiponectin and sphingolipids. Frontiers in Endocrinology (Lausanne). 2016;**7**(33):1-11

[59] Monteiro JP, Maciel E, Melo T, Flanagan C, Urbani N, Neves J, et al. The plasma phospholipidome of *Tursiops truncatus*: From physiological insight to the design of prospective tools for managed cetacean monitorization. Lipids. 2021;**56**(5):461-473

[60] Bories P, Rikardsen AH, Leonards P, Fisk AT, Tartu S, Vogel EF, et al. A deep dive into fat: Investigating blubber lipidomic fingerprint of killer whales and humpback whales in northern Norway. Ecology and Evolution. 2021;**11**:6716-6729

Chapter 4

Typical Changes in Carbon and Nitrogen Stable Isotope Ratios and Mercury Concentration during the Lactation of Marine Mammals

Tetsuya Endo and Mari Kobayashi

Abstract

The increase and decrease in the $\delta^{15}N$ values of offspring owing to the suckling of $\delta^{15}N$-enriched milk (nursing) and the feeding shift from milk to solid food (weaning), respectively, are thought to be common traits observed in mammals. However, there are a few studies on lactation in marine mammals, especially large whales, because samples of calf, lactating mother, and milk are difficult to obtain. In this chapter, we review the studies on reproduction of marine mammals using $\delta^{13}C$ and $\delta^{15}N$ values analyzed in several tissues and describe the typical changes reported to date in those values and Hg concentrations in offspring and milk during lactation. Next, we present data on ontogenetic changes in $\delta^{15}N$ and $\delta^{13}C$ profiles and Hg concentration, especially focusing on the lactation period, in muscle samples of hunted bowhead whale, and stranded common minke whale (mysticetes), Dall's porpoise (odontocete), and the harbor seal (phocid). Finally, we compare the $\delta^{15}N$ and $\delta^{13}C$ values in muscle samples of calves from common mink whale, Dall's porpoise, and killer whale and suggest that these values could be excellent proxies for maternal forging habits and trophic levels.

Keywords: lactation, calf, pup, marine mammal, stable isotope of ratios, mercury

1. Introduction

Viviparity and lactation are the most important traits in mammals. The developing fetus is connected to the placenta via an umbilical cord and obtains nutrients through it. Milk is the most complete and natural food of mammals that offspring can consume during the early stages of their life, as it ensures proper nutrition and development. Two types of reproductive taxa have been proposed in mammals based on whether a pregnant and lactating female can catabolize the energy stored in the body to grow fetuses and produce milk [1–4]. Mammals who catabolize accumulated energy resources are called capital breeders, whereas income breeders use energy resources gained concurrently; mysticetes and true seals are generally classified as capital breeders, whereas odontocetes and fur seals are generally classified as income breeders.

IntechOpen

Cetaceans give birth and suckle in inaccessible oceans, making it difficult to observe their reproduction, especially in large mysticete whales. Studies focused on the biochemistry and ecology of large whales have been mainly conducted using preserved samples from commercial whaling in the past, but there is little information on reproduction because of the restricted whaling for lactating females and calves. Owing to the international ban on whaling, new samples from large cetaceans are not easy to collect [5–7]. Currently, most studies on lactation and mother-to-offspring relations in marine mammals are conducted using pinnipeds as they give birth and nurse pups on land or ice in accessible areas; they are easy to observe and sample [4, 6–9].

Stable isotope ratios of nitrogen ($\delta^{15}N$) and carbon ($\delta^{13}C$) are used exclusively to study the ecology and biochemistry of marine mammals, and most samples used in these studies are the muscle, liver, blood components, hair, whiskers, bones, skin, and teeth. However, the integrated terms of dietary information are different among these tissues because of different turnover rates [10, 11]. Some studies analyzed two tissues at different turnover rates from individual animals, that is, red blood cells and plasma samples, bone and muscle samples, and hair and plasma samples, to estimate the past and recent information on dietary ecology and biochemistry [12–16].

As the contamination of mercury (Hg) and cadmium (Cd) in milk is very low [17], these burdens in offspring are usually low. However, after weaning, Hg and Cd concentrations in some mammals tend to increase sharply, reflecting the feeding on fish and cephalopods [18–22]. In contrast, adult marine mammals located at high trophic positions are extremely contaminated with Hg [23, 24]. As Hg passes through the placenta, there is great concern regarding the neurotoxicological effects of Hg on infants of not only human populations but also marine mammals [25, 26]. As fetal samples of marine mammals are difficult to obtain, the Hg exposure of the fetus is indirectly estimated using the Hg concentrations in lanugo hair [16, 27–29] and in the red blood cells of neonates [16].

This chapter first describes the different reproduction strategies between mysticetes and odontocetes and between seals and fur seals, and then reviews the studies on lactation in marine mammals using the $\delta^{15}N$ and $\delta^{13}C$ signatures and Hg concentration. Next, we discuss our data on the ontogenetic profiles (focused on lactation) of $\delta^{15}N$ and $\delta^{13}C$ signatures and Hg concentration in muscle samples of stranded common minke whale (*Balaenoptera acutorostrata*; MW) (mysticete), Dall's porpoise (*Phocoenoides dalli*; DP) (odontocete), and Kurill harbor seals (*Phoca vitulina stejnegeri*, HS) (pinniped) analyzed in our laboratory. Finally, we compare the $\delta^{15}N$ and $\delta^{13}C$ values in muscle samples of stranded calves from MW, DP, and killer whale (*Orcinus orca*; KW), and suggest that these values in calves could be excellent proxies for maternal forging habits and trophic levels.

2. Lactation in marine mammals

2.1 Lactation and milk composition

Most baleen whales (mysticetes) are characterized by long migrations between the feeding and breeding grounds, and the storage of adequate energy for pregnancy and lactation, giving birth and lactating but fasting or eating relatively little [6]. Most baleen whale species have relatively brief lactation (nursing and weaning) periods of 5–7 months, except for bowhead whale (*Balaena mysticetus*, BW) [5, 6]. They produce milk relatively low in water (40–50%), high in fat (lipids) (30–50%), and moderately high in protein (9–15%), relying on energy stored before parturition to support the rapid fattening and growth of their calves over a brief lactating period [6]. In contrast, most toothed whales, dolphins, and porpoises (odontocetes) have much more extensive lactations typically lasting 1–3 years, during which the mothers feed. Their milk is higher in water (60–70%) and lower in fat (10–30%) than that of mysticetes but has similar levels of protein (8–11%) [6]. Cetaceans with brief lactation periods tend to produce high-fat milk, investing heavily in offspring for a brief period compared to those with prolonged lactation periods, which tend to produce low-fat milk. The prolonged weaning period in odontocetes may be related to the learning of predatory and social skills, whereas mysticetes, which feed on swarms of zooplankton and schools of small fish, do not need such learning.

The mysticete-odontocete contrast is similar to the contrast between the true seals (family Phocidae) and the fur seal (family Otariidae) [4, 6–8, 30], although both families are piscivores. Most phocid mothers fast during the lactation period rely on energy stored before parturition, which is generally less than 2 months in duration, and pups are usually weaned abruptly. In contrast, otariid mothers feed throughout lactation (except for the initial perinatal period), which may last 1–2 years, and most otariid species are weaned gradually. Furthermore, the milk of phocids is more energy-rich (higher fat concentration) than that of otariids. The different reproduction strategies between mysticetes and odontocetes and between phocids and otariids can be explained by the different breeding strategies as capital and income breeders [1–4].

Weaning involves the transition from nutritional dependence on milk to solid food, and milk and solid food can often be found in the stomachs of gradually weaned offspring. Body lengths (BL) in weaned cetaceans have been reported to be correlated with maternal BL [31].

2.2 Carbon and nitrogen stable isotope ratios

Stable isotope analyses of carbon ($\delta^{13}C$) and nitrogen ($\delta^{15}N$) are useful tools for obtaining information on feeding ecology. The $\delta^{15}N$ value shows a stepwise increase with increases in the trophic level through a food chain, whereas the $\delta^{13}C$ value is used to estimate the relative contribution to the diet of potential primary sources [11, 32, 33]. Significant increases of $\sim$3‰ in the $\delta^{15}N$ value have been shown to occur between predators and their prey, whereas the trophic fractionation for $\delta^{13}C$ values is smaller than that for $\delta^{15}N$ values, averaging approximately 1‰ [11, 32–35] (**Figure 1**). The $\delta^{13}C$ and $\delta^{15}N$ profiles are used in studies of habitat preferences such as pelagic *vs.* benthic, and nearshore *vs.* offshore *vs.* estuarine [33], as well as the geographical differences of inhabitants [11, 32, 36, 37]. Furthermore, mother-to-offspring transfer of nutrients owing to lactation in marine mammals has been preferentially investigated using $\delta^{13}C$ and $\delta^{15}N$ signatures. Higher trophic levels ($\delta^{15}N$ values) in nursing offspring than those in their mothers are widely observed in terrestrial and marine mammals [3, 11, 32, 33, 38–40], as the trophic level of milk is higher than that of foods which the mother feeds on (**Figure 1**). In contrast, the $\delta^{13}C$ value of offspring does not show a clear-cut pattern because of the large variation in ^{13}C-depleted lipids (fat) concentration in the milk they suckled [3, 33, 40].

As the $\delta^{13}C$ values of lipids in mammals are typically 5–7% lower than those of tissue proteins or carbohydrates [11], most studies have analyzed lipid-extracted tissues to remove the potential confounding effect of variation in tissue lipid content on the $\delta^{13}C$ values [11, 32, 41]. Consistent with this, the $\delta^{13}C$ values in lipid-extracted

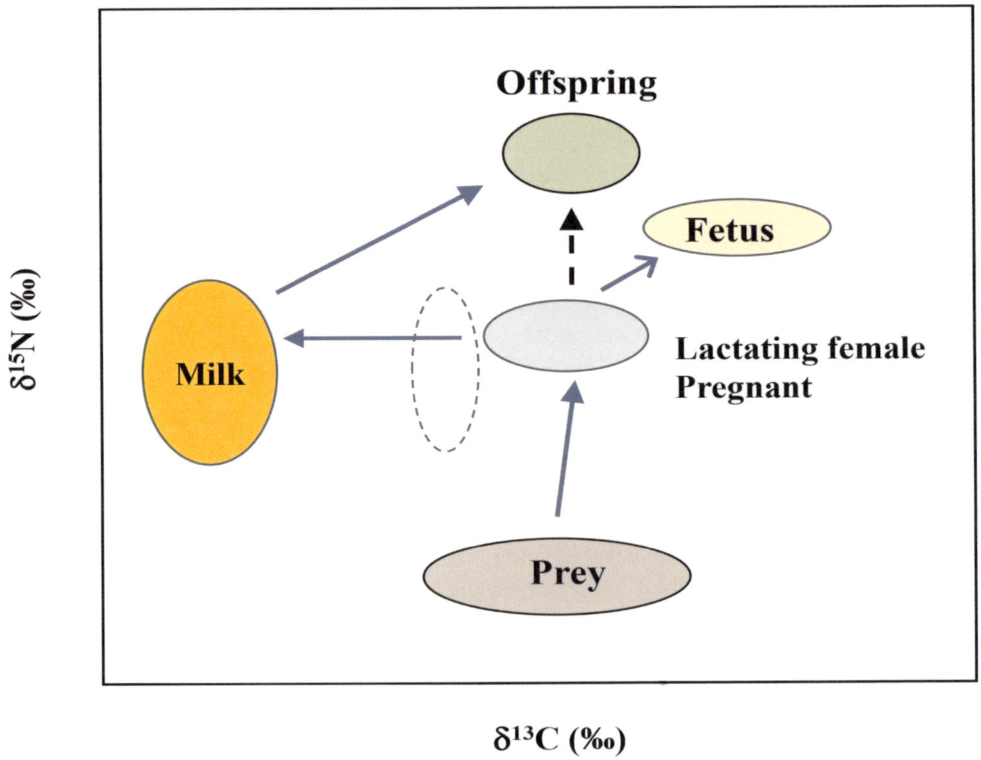

Figure 1.
Diagram showing the relationships among lactating female, milk and offspring and between pregnant and fetus by stable isotope ratios of carbon and nitrogen. The dashed ellipse indicates the lipid extracted milk.

milk (shown as a dotted ellipse of milk protein in **Figure 1**) were approximately 5–6% higher than those in milk containing lipids [3, 9, 42, 43].

2.3 Isotopic discrimination and half-life in tissues

As mentioned above, the enrichment of $\delta^{15}N$ values between whole predator and prey ($\Delta^{15}N_{predator\text{-}prey}$) is ~3‰ [3, 11, 32–35]. However, it is practically impossible to measure the whole body $\delta^{15}N$ value in large animals; the $\delta^{15}N$ value in a tissue is usually substituted for the $\delta^{15}N$ value of the whole body, despite the $\delta^{15}N$ value in each tissue is different. For instance, the highest $\delta^{15}N$ value among seven tissues of fin whales (*Balaenoptera physalus*; FW) was the brain (11.4 ± 0.36%), the middle was muscle (9.88 ± 0.58%), and the lowest was the bone protein (9.19 ± 0.71%) of with a difference of ~2.2% [44]. The following is the order of $\delta^{15}N$ values in tissues that are widely observed [34, 35, 39, 44–46].

> *brain > plasma (serum) > liver > muscle = hair (fur) = skin = baleen > red blood cells.*

Thus, the $\Delta^{15}N_{predator\text{-}prey}$ values calculated for each tissue are different. Among these tissues, the $\delta^{15}N$ value of muscle tissue may closely reflect that of the whole body, as the mass of muscle tissue is the highest in the body (for example, ~30% in humans) [11, 32].

Tissue turnover rates are important for understanding the timeframe of dietary information preserved by the $\delta^{15}N$ and $\delta^{13}C$ signatures in a given tissue. Tissue turnover should be closely linked to protein turnover, as the tissues typically used for isotopic analyses (hair, muscle, plasma/serum, red blood cells, bone collagen, etc.) primarily comprise proteins, with protein turnover being the most rapid in the liver and plasma/serum, followed by that in the muscle, and slow in red blood cells and bone collagen [11, 32, 47].

The tissue turnover rate in animals generally decreases with body size; for example, the time until 50% replacement of red blood cells was 35 days in humans, but 114 days in beef cattles [11]. Furthermore, the tissue turnover rate in animals is rapid in growing animals, as also in metabolically active tissues [48, 49]. Therefore, as we previously implied, the turnover rate of muscle in MW calves may be markedly faster than that in MS adults, with little time lag between the feeding shift from milk to solid food and the change in $\delta^{15}N$ signature in the calf muscle [40].

Unlike the blood, liver, and kidney, the hair, baleen plates, and whiskers do not undergo turnover; past information is preserved without being erased. Hair and baleen plates are reported to grow continuously, at a rate of approximately 1 cm/month for humans [11] and 12.9–20 cm/year for Balaenopteridae species [50], respectively; thus, sampling close to the roots of hair, baleen plates, and whiskers will record the more recent information ($\delta^{13}C$ and $\delta^{15}N$ values), whereas sampling farther along those will record the information in the past [1, 51–54]. In contrast, the dentin growth layer in teeth allows us to investigate the annual changes in $\delta^{13}C$ and $\delta^{15}N$ signatures in beluga whales (*Delphinapterus leucas*) [55], KW, and sperm whales (*Physeter microcephalus*) [33]. Studies using the dentin layer have reported a decrease in the $\delta^{15}N$ profile because of weaning, with little information on nursing that occurs in a brief period of less than one year. Aubail et al. [56] analyzed not only the $\delta^{13}C$ and $\delta^{15}N$ values but also the Hg concentration in the dentin layers of ringed seals (*Phoca hispida*) and reported a high Hg concentration owing to the placental transfer of Hg and high $\delta^{15}N$ values owing to nursing in the first layer (first year) of dentin.

Bones are well-preserved samples, but they can only be obtained from cadavers. Because of their slow turnover rate, it is suitable for the study of prolonged lactation in animals. Jansen et al. [14] and Vales et al. [57] investigated the ontogenetic dietary changes in harbor porpoises (*Phocoena phocoena*) and South American fur seals (*Arctocephalus australis*) using bone samples. However, the time lag between the weaning and the $\delta^{15}N$ signature in the bone should be considered when analyzing the results.

Most blood and milk samples used for the studies on lactation and feeding ecology of marine mammals were obtained from pinnipeds because they are easy to handle and can be sampled blood from pup and mother pairs, and milk and pup tissue pairs [9, 13, 43], in addition to lanugo hair and whisker samples, which is not possible in large whales.

2.4 Isotopic relationships among lactating female, milk, and offspring and between pregnant female and fetus: strategies of income and capital breeders

Figure 2 shows the $\delta^{15}N$ and $\delta^{13}C$ values in the muscle samples of KWs that were mass-stranded in Hokkaido. Nine corpses, including three lactating mothers and three calves at few months in age, (including two pairs of lactating mothers and calves), were recovered [58, 59]. Stranded lactating mothers could starve for several days before death, whereas milk was found in the stomach of the calf corpses.

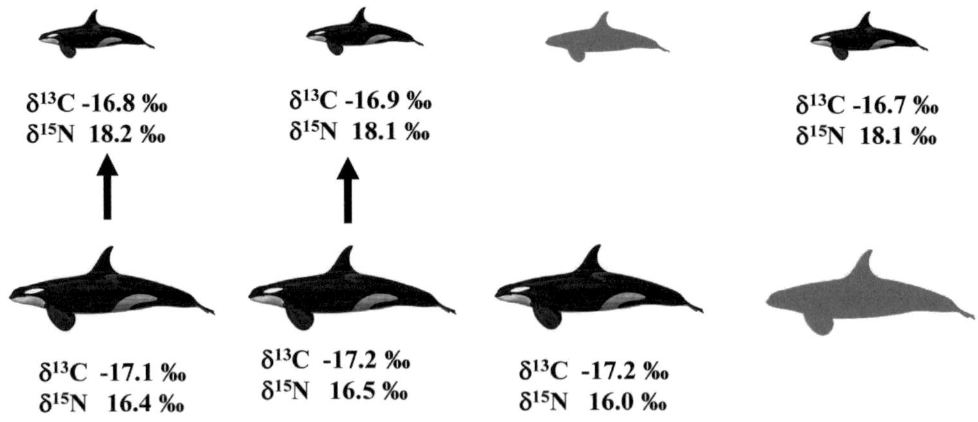

Figure 2.
Stable isotope ratios of carbon and nitrogen in muscle of killer whales from 3 calves and 3 lactating females mass-stranded in Hokkaido. Genetic analysis revealed two pairs of mother-calf relationship. See ***Table 1.***

The $\delta^{15}N$ and $\delta^{13}C$ values in the muscle samples were similar between the three calves and the three lactating mothers. The $\Delta^{15}N_{offspring\text{-}mother}$ values of two pairs of whales were 1.7‰ and 1.6‰, and the $\Delta^{13}C_{offspring\text{-}mother}$ values of two pairs were slightly positive 0.3‰ and 0.3‰ (**Figure 2**). To the best of our knowledge, the $\Delta^{15}N_{offspring\text{-}mother}$ values in muscle samples (1.7‰ and 1.6‰) have not yet been reported in cetaceans within nature; however, these values were consistent with those calculated from blood and milk samples of pinniped pairs [9, 39, 60] and captive mammalian pairs [61]. $\Delta^{15}N_{offspring\text{-}mother}$ in KWs (1.7‰ and 1.6‰) were smaller than those reported for $\Delta^{15}N_{predator\text{-}prey}$ (~3‰), as the lactation of KW calves was at an early stage of nursing before the $\delta^{15}N$-enriched peak (**Figures 3–5**).

In contrast, the $\Delta^{15}N_{offspring\text{-}mother}$ values in the liver samples were slightly negative (−0.5‰ and − 0.5‰) (**Table 1**). These phenomena may be explained by the faster

	Remark	Muscle			Liver			Blubber	
		$\delta^{13}C$ (‰)	$\delta^{15}N$ (‰)	Hg[a]	$\delta^{13}C$ (‰)	$\delta^{15}N$ (‰)	Hg[a]	PCB[b, c]	*p,p'*-DDE[c]
AKW6	Lactating female	−17.1	16.4	1.27	−16.8	19.2	38.0	42.3	109
AKW7	Calf of AKW6	−16.8	18.2	0.07	−16.6	18.7	0.30	68.2	237
AKW9	Lactating female	−17.2	16.5	1.26	−16.7	19.2	62.4	26.2	50.5
AKW8	Calf of AKW9	−16.9	18.1	0.08	−16.7	18.7	0.50	42.3	98.9
AKW2	Lactating female (no calf among AKW pod)	−17.2	16.0	1.26	−16.8	18.2	57.4	31.4	68.6
AKW3	Calf (no mother among AKW pod)	−16.7	18.1	0.10	−16.6	18.6	0.30	41.6	112

[a]*total mercury concentration (μg/wet g).*
[b]*sum of 12 PCB isomers.*
[c]*(mg/lipid wet g).*
From Endo et al. [58], and Haraguchi et al. [62].

Table 1.
Analytical results of stable isotope ratios of carbon and nitrogen and mercury concentrations in muscle and liver and PCB and p,p'*-DDE concentrations in blubber of killer whales mass-stranded in Hokkadio (AKW pod).*

turnover rate of nitrogen in the liver than in the muscle [10, 47], and the increase in maternal $\delta^{15}N$ values by starvation [63]. The Hg concentrations in the muscle and liver samples of lactating mothers were markedly higher than those of the calves, and this trend was particularly pronounced in the liver samples. In contrast to Hg, polychlorinated biphenyl (PCB), and 1,1-dichloro-2,2-bis(p-chlorophenyl) ethylene (*p,p*'-DEE) concentrations in blubber samples were significantly higher in calves than in lactating mothers because of the suckling of contaminated milk containing lipophilic compounds of PCB and *p,p*'-DEE [62].

Two types of reproductive strategies have been proposed for mammals [1–4, 9]. In capital breeders, lactating mothers catabolize their tissues to grow fetuses and produce milk. The trophic levels ($\delta^{15}N$ values) of fetuses and milk are higher than those of pregnant females and lactating mothers, respectively (**Figure 1**); that is, the fetus and nursing offspring consume their mother's tissues [32, 33]. In contrast, in income breeders, fetuses and milk should not be enriched in ^{15}N during parturition and lactation, as resources ingested by the pregnant female and lactating mother will not have been incorporated into her own tissues, and these are directly routed into fetal tissue and milk production.

Considering nitrogen in more detail, capital breeders show positive $\Delta^{15}N_{milk\text{-}mother}$ values: 2‰ in the northern elephant seal (*Mirounga angustirostris*) [13], 1‰ in polar bears (*Ursus maritimus*) [42], 0.5‰ in the southern right whale (*Eubalaena australis*) [38], and 0.3‰ in FW [3] (**Figure 1**). Contrary to the positive values in capital breeders, the $\Delta^{15}N_{milk\text{-}mother}$ values of income breeders are negative: -1.0 ± 0.5‰ in nine income-breeder species [61], -0.8‰ in Steller sea lion (*Eumetopias jubatus*) [43] -0.3‰ in deer mice (*Peromyscus maniculatus*) [2], and $-$ 1.61‰ and $-$ 1.48‰ in two species of fur seals (*Arctocephalus gazella* and *Arctocephalus tropicalis*) [9]. However, Chilvers [64] recently reported a positive $\Delta^{15}N_{milk\text{-}mother}$ value of 0.4‰ in the New Zealand sea lion (*Phocarctos hookeri*), an income breeder, and suggested that the production of milk is not only from the diet but also through catabolizing of tissue. In contrast to milk, sampling of fetuses from marine mammals, especially cetaceans, is very rare, with only a single $\Delta^{15}N_{fetus\text{-}mother}$ value of 1.6‰ reported in FW, a capital breeder [3].

2.5 Mercury and cadmium

As apex predators, odontocetes are exposed to high levels of pollutants, such as Hg and organochlorines, through feeding on contaminated fish [65–67]. In contrast, mysticetes, which feed lower on the food chain (zooplankton and small fish), typically have less significant exposure to pollutants. Cd is accumulates in molluscs, particularly cephalopods, and accumulates in predators via not only marine food webs but also species-specific physiologic mechanisms [18, 19, 68]. Cd is likely to be distributed preferentially in the kidney rather than in the liver of marine mammals, whereas Hg tends to be distributed preferentially in the liver [18, 19, 21, 58]. A sharp increase in the Hg burden in the liver is thought to be a weaning proxy for fish eaters in marine mammals, whereas a sharp increase in the Cd burden in the kidney could be a weaning proxy for cephalopod eaters [22]. The degree of Hg burden in mysticetes is generally low, but the Hg burden in opportunistic feeders of mysticetes could reflect the amount of fish consumed [69].

Methyl Hg, produced within marine ecosystems, is a neurotoxicant that is transported across the blood-brain barrier and placental barrier [26]. There is great concern regarding the neurotoxicological effects of Hg on infants of the human population and piscivores [26]. Because of the difficulty of sampling from fetal tissue, Hg exposure in the fetus is usually estimated using the Hg concentrations in the scalp hair of infants [26], lanugo hair [27–29], dentin [56], and red blood cells of neonates [16]. Rea et al. [28] reported a high level of Hg (> 40 μg/g) as well as $\delta^{15}N$ and $\delta^{13}C$ values in the fur of sea lion pups, suggesting that their mothers ate fish contaminated with Hg at the high trophic position during the late gestation. Furthermore, the Hg concentration in red blood cells of suckling offspring could reflect the Hg burden in their fetus stage [26], as the majority of red blood cells are likely to be produced during late gestation [13, 61]. As hair is an excretory route for Hg, the Hg in hair is used for biomonitoring of Hg burden in human populations as well as pinnipeds, and in humans, scalp hair Hg is on average 250 times greater than that in whale blood [11]. Endo et al. [70] reported high Hg concentrations in the scalp hair of residents living in a whaling town and a positive correlation between the Hg concentrations and the $\delta^{15}N$ values in their hair. Brookens et al. [71] and Hobson et al. [52] reported high Hg concentrations in the hair of Pacific harbor seals (*Phoca vitulina richardii*) and in the baleen plate of MWs and the correlations between Hg concentrations and $\delta^{15}N$ values in these samples.

To the best of our knowledge, the only study on Hg analyzed in the fetus of marine mammals was conducted by Itano et al. [23] using hunted striped dolphins (*Stenella coeruleoalba*), which is heavily contaminated with Hg among odontocetes inhabiting waters around Japan [24]. Itano et al. [23] reported an increase in the Hg burden in striped dolphin fetuses with the increase in pregnancy term. Comparing muscle Hg concentrations in fetuses, suckling offspring, and mature females, they reported values of 0.900 ± 0.310, 0.520 ± 0.060, and 11.100 ± 0.300 μg/wet g, respectively. In addition, they reported trace levels of Hg in milk (26 ng/g) and high levels of Hg in the blood of adult females (1.200 ± 0.200 μg/g), which was 46 times higher than that in the milk. Thus, the Hg concentration in calves, which are contaminated with a high Hg burden through the placenta, is likely to be diluted by the growth dilution effect, even though a small amount of Hg could be taken from milk. In contrast, the Hg concentration in breast milk and blood in women at 6 weeks after delivery were 0.6 ± 0.4 and 2.2 ± 1.9 ng/mL, respectively, and that in blood was only ~3 times higher [25]. No further studies on Hg transfer through the placenta and milk have been conducted in cetaceans.

2.6 Ontogenetic changes in $\delta^{13}C$ and $\delta^{15}N$ values in bowhead whales

Large-scale ontogenetic studies on baleen whales, focusing on lactation using $\delta^{13}C$ and $\delta^{15}N$ signatures, have only been conducted in BW samples from native subsistence hunts [20, 72, 73]. The BWs are large baleen whales inhabiting the icy Arctic waters, and they are born at 4–5 m BL, weaned at ~10 m BL, with sexual maturity at over 13.5 m BL [72, 74]. BWs feed on pelagic zooplankton and benthic amphipods, where the $\delta^{13}C$ and Cd levels of benthic amphipods are lower and higher than those of zooplankton, respectively, and Hg is in trace concentrations [20].

Figure 3 shows the ontogenetic changes in $\delta^{15}N$ and $\delta^{13}C$ signatures in the muscle samples of BWs by Lee et al. [72]. They analyzed the $\delta^{15}N$ and $\delta^{13}C$ values of non-lipid–extracted muscle samples. Line smoothing shows the $\delta^{15}N$-enriched peak at ~8 m BL and a $\delta^{15}N$-depleted peak at ~10 m BL. A large study on BWs conducted thereafter by Horstmann-Dehn et al. [73] showed prominent $\delta^{15}N$-enriched peaks in muscle (n = 133) and epidermis (n = 130) samples of BWs at ~8 m BL. The increase in $\delta^{15}N$ values to the enriched peak (~8 m BL) could represent nursing, and this peak is likely to be the onset of weaning (shift from milk to solid foods). Furthermore, the BL

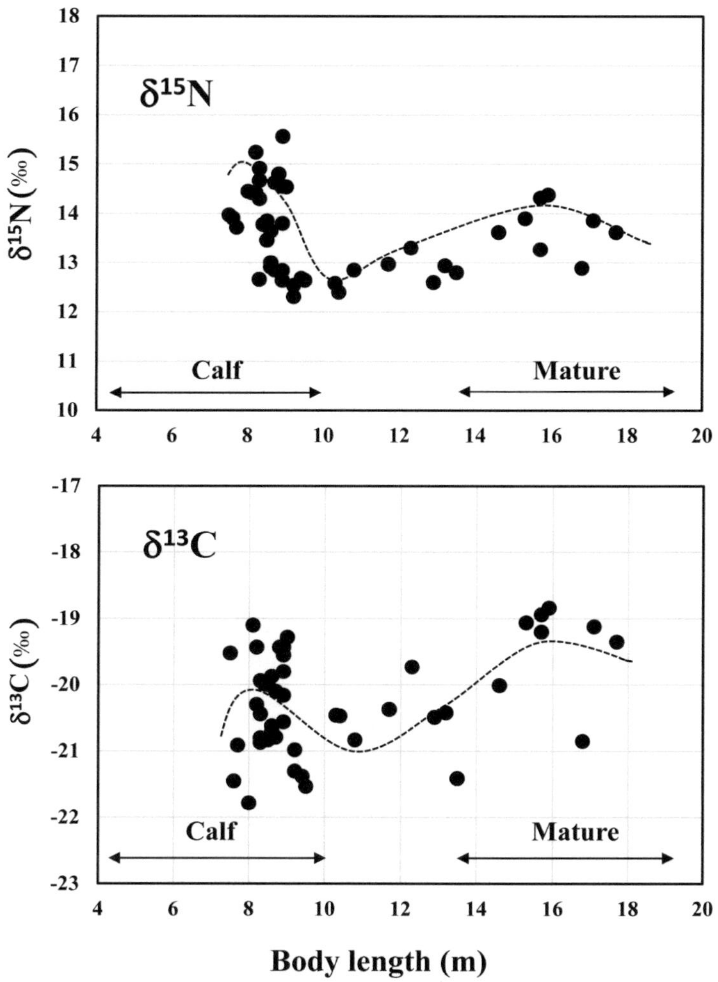

Figure 3.
Ontogenetic changes of δ^{15} and δ^{13}C signatures in muscle of bowhead whales. These figures were drawn based on the data from Table 5 (n = 47) reported by Lee et al. [72] with permission. Two outliers were not drawn.

at the δ^{15}N-depleted peak may represent BL at complete weaning in BWs, which coincides with the weaned BL at ~10 m reported previously [72, 74]. After weaning, the δ^{15}N values gradually increased and reached a plateau. The ontogenetic change in the δ^{13}C profile was similar but more prominent than that of δ^{15}N change; no correlation was found between the δ^{13}C and δ^{15}N values. The prominent increase in post-weaning δ^{13}C values may be related to the feeding on zooplankton and benthic amphipods, although the cofounding effect of tissue lipid content on δ^{13}C values should be considered: large differences are found between the lipid-extracted and non-extracted muscle samples when δ^{13}C values are low [73]. Dehn et al. [20] reported trace levels of Hg in the muscle (0.02 ± 0.01 μg/wet g) and a moderate level of Cd in the kidney (15.08 ± 14.94 μg/g) of BWs, and these levels of Hg and Cd could be consistent with their habit of feeding on pelagic zooplankton and benthic amphipods. Despite a large number of BW samples, Lee et al. [72] and Horstmann-Dehn et al. [73] did not investigate sex-related differences in the δ^{13}C and δ^{15}N profiles of BW samples.

Dehn et al. [20] reported a marked depletion in $\delta^{13}C$ values (between −24‰ and − 25‰) in two fetus of BWs at ~1 m and ~ 4 m BL as compared to those in adults (approximately −20‰), and these values are in the range of the $\delta^{13}C$ values in milk found widely in marine mammals (between −24‰ and − 26‰) [3, 9, 43, 64]. Dehn et al. [20] did not mention the $\delta^{15}N$-enrichment in the muscle of BW fetuses, but Horstmann-Dehn et al. [73] reported the enriched $\delta^{15}N$ values in the epidermis of BW fetuses. Borrell et al. [3] reported higher $\delta^{13}C$ and $\delta^{15}N$ values in FW fetuses (−17.45 ± 0.53‰ and 11.46 ± 0.38‰, n = 10) than in lactating females (−18.06 ± 0.22‰, and 9.46 ± 0.55‰, n = 13), although they did not mention the BL of the analyzed fetuses. More research on $\Delta^{13}C_{fetus\text{-}mother}$ values in addition to $\Delta^{15}N_{fetus\text{-}mother}$ values is needed to clarify the transport of nutrition from mother to fetus. The BW and FW are mysticetes and are classified as capital breeders.

3. Ontogenetic change in $\delta^{15}N$ and $\delta^{13}C$ signatures of marine mammals stranded in Hokkaido, Japan

3.1 Stranding of marine mammals along the coast of Hokkaido, Japan

Hokkaido is the northernmost island in Japan and is surrounded by the North Pacific Ocean, the Sea of Japan, and the Sea of Okhotsk. Most stranded cetaceans in Hokkaido are odontocetes, such as DP, harbor porpoise (*Phocoena phocoena*), Pacific white-sided dolphin (*Lagenorhynchus obliquidens*), and the mysticete MW. In this section, we describe the $\delta^{15}N$ and $\delta^{13}C$ profiles in muscle samples of MW and DP calves, in addition to the muscle samples of Kurill harbor seals (*Phoca vitulina stejnegeri*; HS) stranded in Hokkaido. Furthermore, we compared the $\delta^{15}N$ and $\delta^{13}C$ values in calf muscle samples from MW, DP and KW stranded in Hokkaido. We analyzed the $\delta^{15}N$ and $\delta^{13}C$ values in the muscle samples after lipid extraction.

3.2 Ontogenetic changes in $\delta^{15}N$ and $\delta^{13}C$ values in common minke whales

Figure 4 shows the $\delta^{15}N$ and $\delta^{13}C$ signatures with a focus on calves of MW [40]. MWs are mysticete and opportunistic feeders that change their prey items temporally and regionally; they feed on zooplankton and small fish. MW may strand during their migration [75].

A clear $\delta^{15}N$-enriched peak was found in MW calves (n = 12), including newborn animals at 2.6 and 3.0 m BL. Similar to BWs (**Figure 3**), the increase in $\delta^{15}N$ values toward this peak could represent nursing, and the decrease in $\delta^{15}N$ values from this peak could represent weaning [40]. The $\delta^{15}N$ values of calves fitted to a quadratic function ($p < 0.05$), and this peak calculated by the fitted equation was 4.0 m BL and 13.3‰, suggesting the onset weaning at ~4 m BL. In addition, the BL of weaned animals was estimated at ~5 m BL from the fitted equation and the average $\delta^{15}N$ values of mature animals as 11.9‰, which coincided with the reported BL of weaning [76]. Thus, the weaned BL of MWs (**Figure 4**) and BWs (**Figure 3**) estimated using $\delta^{15}N$ profiles in muscle samples coincided with the weaned BL estimated from ecological and morphological studies. We believe that the $\delta^{15}N$ value of the muscle from calves of mysticetes reflects the dietary shift quite rapidly with little time lag, because of the brief weaning period [5, 6] and the fast turnover rate of small and growing animals [48, 49].

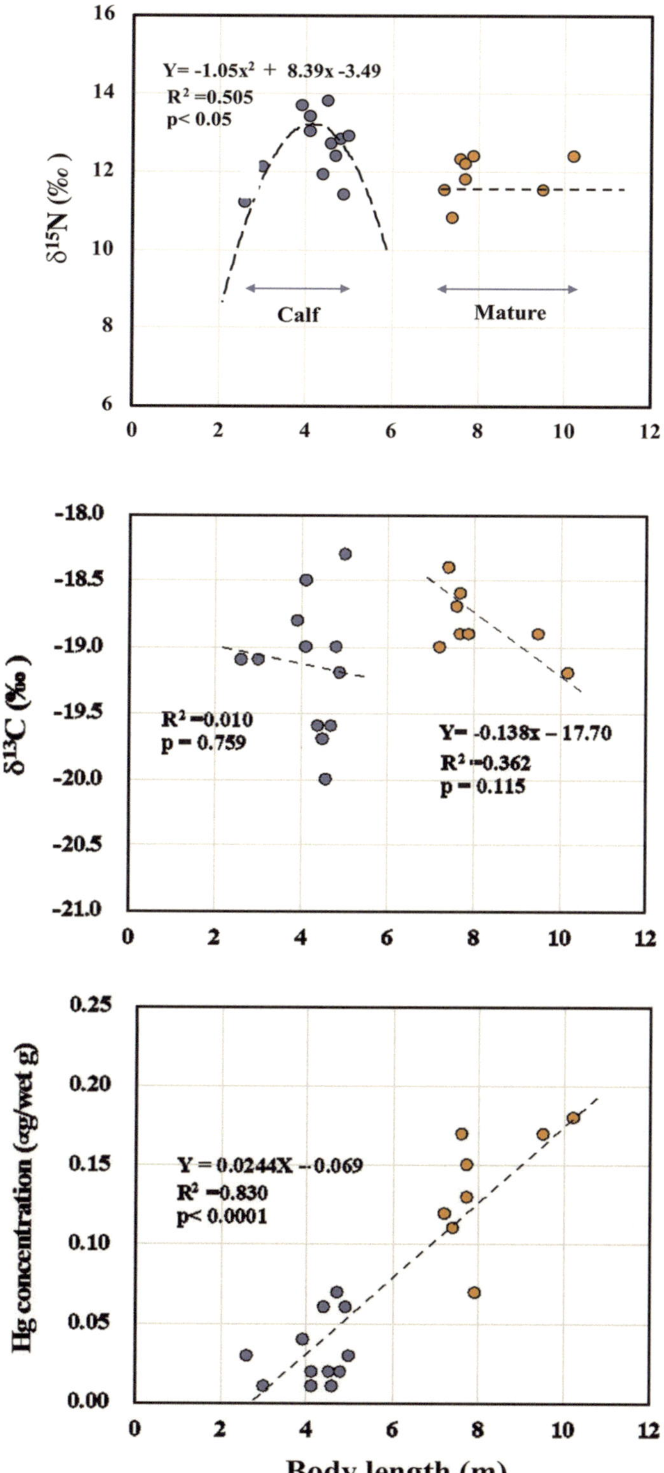

Figure 4.
Ontogenetic changes of $\delta^{15}N$ and $\delta^{13}C$ signatures in muscle of common minke whale stranded in Hokkaido. This figure was reprinted from Endo et al. [40] with permission from Aquatic Mammals.

No particular pattern was found in $\delta^{13}C$ values of MW calves, probably owing to the large variation in $\delta^{13}C$-depleted lipid concentrations in milk [3, 33, 42].

The Hg concentrations in immature (weaned) animals were apparently higher than those in MW calves because of the opportunistic eating of fish. Marked increases in Hg owing to weaning were found in liver samples of MWs (data not shown).

Studies on the ontogenetic changes in $\delta^{13}C$ and $\delta^{15}N$ values in mysticetes were conducted by Borrell et al. [3] using muscle samples from FWs and by Mitani et al. [53] using baleen plate samples of MWs. However, the study on FWs did not include calf samples, and a clear $\delta^{15}N$-depleted peak related to weaning was not found among the few $\delta^{15}N$-depleted peaks observed in the baleen plate of MWs.

3.3 Ontogenetic changes in $\delta^{13}C$ and $\delta^{15}N$ values in *Dalli*-type Dall's porpoises

Dalli-type DP (odontocete) is widely distributed in the northern part of the North Pacific and move from south to north seasonally in the North Pacific Ocean [75]. According to Kasuya [77], the calves are born from August to September at 100 cm BL and nursed for approximately 2 years. In addition, sexual maturity is attained in males at the age of 7.9 years at 195.7 cm BL and in females at 6.8 years at 186.5 cm BL. Huang et al. [31] estimated the weaned BL of DPs to be 135 cm. We analyzed the $\delta^{15}N$ and $\delta^{13}C$ values and Hg concentrations of muscle samples of DPs stranded along the coast of Hokkaido.

An $\delta^{15}N$-enriched peak at ~115 cm BL owing to lactation was found in the muscle samples of DPs (**Figure 5**), similar to that observed in the mysticetes BWs and MWs (**Figures 3** and **4**). The $\delta^{15}N$ values following the peak gradually declined at ~160 cm BL, then increased slightly and plateaued. The BL at the $\delta^{15}N$-enriched peak could imply the BL at the onset of weaning, although available information on the onset of weaning in DPs is lacking. The $\delta^{15}N$-depleted peak at ~160 cm BL may be related to BL at complete weaning, although the reported BL of DP at weaning DP is 135 cm [31]. The DP sample shown in **Figure 5** was biased and did not include the DP samples at 130–150 cm BL at which weaning was reported to occur.

Although the $\delta^{13}C$ values varied considerably, small $\delta^{13}C$-enriched and $\delta^{13}C$-depleted peaks were found at ~130 cm BL and ~ 190 cm BL, respectively, and the small $\delta^{13}C$-enriched peaks may be related to weaning. No correlation was found between the $\delta^{13}C$ and $\delta^{15}N$ values of DPs, similar to that of the mysticetes (**Figures 3** and **4**). We believe that the large variability in $\delta^{15}N$ and $\delta^{13}C$ values between 180 and 230 cm BL may be due to sex-related differences, but this was not clear (males and females are not shown separately in **Figure 5**). Ontogenetic changes in $\delta^{15}N$ and $\delta^{13}C$ profiles in DP samples (**Figure 5**) were similar to those in BW samples (**Figure 3**).

The Hg concentrations in DPs increased with increases in BL, and were markedly higher than those of MWs (**Figure 4**), reflecting the higher trophic position of DPs.

The ontogenetic signatures of $\delta^{15}N$ and $\delta^{13}C$ in KWs were investigated by Newsome et al. [33] using the dentin growth layer of teeth. They reported a decrease in $\delta^{15}N$ values related to weaning in those dentins with no $\delta^{15}N$-enriched peaks related to nursing, and a gradual and continuous increase in $\delta^{15}N$ values after weaning. This post-weaning increase in $\delta^{15}N$ values in KWs is more prominent than that in DPs (**Figure 5**), ringed seals [56], and the beluga whales [55], and similar to that in harbor porpoises [14] and South American fur seals [57]. We believe that the increase in $\delta^{15}N$ values after weaning may be correlated with the increase in trophic level of juvenile animals owing to growth.

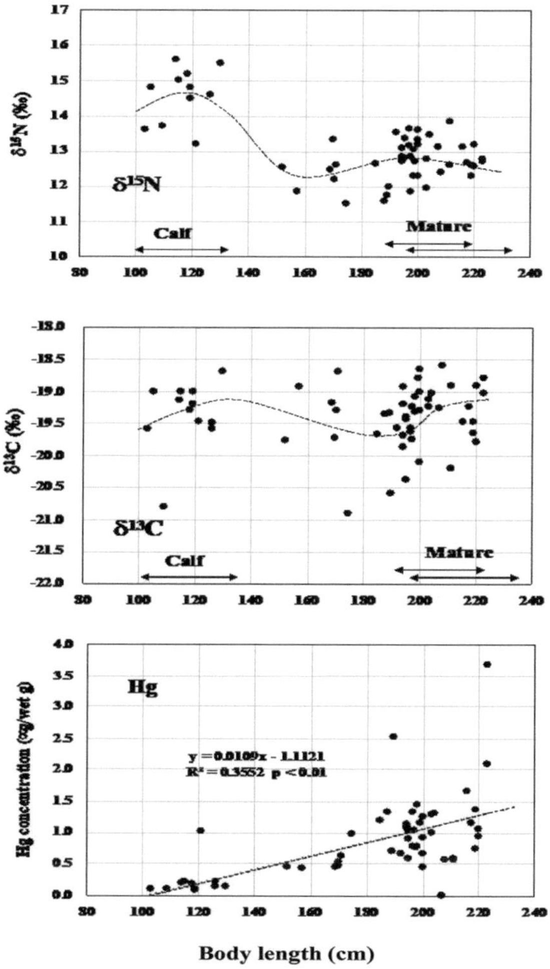

Figure 5.
Ontogenetic changes of δ¹⁵N and δ¹³C signatures and Hg concentration in muscle of Dall's porpoises stranded in Hokkaido, Japan.

3.4 Ontogenetic changes in δ^{13}C and δ^{15}N values in kurill harbor seals

HSs are distributed on the eastern coastline of Hokkaido, Japan. According to Naito and Nishiwaki [78], HSs is are born at 98 cm BL after a 9-month gestation period, and weaning occurs at approximately 4 weeks after birth at ~110 cm BL. It is unknown whether weaning occurs suddenly or gradually, but the weaning is likely rapid, considering the brief lactation period. Males attain sexual maturity at 140 cm BL (4 years) and females at 133 cm BL (3 years), and the maximum BL observed was 191 and 175 cm in males and females, respectively. HSs feed on prey in inshore areas, such as cephalopods, crustaceans, and small fish. The competition between HSs and coastal fisheries has recently become a serious problem in Hokkaido because of the increased number of HSs. Tissue samples were collected from stranded HSs in salmon set nets. **Figure 6** shows our data on HS males and females.

A δ^{15}N-depleted peak was found at 120–130 cm BL of males, and the δ^{15}N values increased thereafter, whereas a small δ^{13}C-depleted peak was found at

120 cm BL, and $\delta^{13}C$ values gradually increased thereafter. In contrast, a $\delta^{15}N$-depleted peak was observed in females, whereas the $\delta^{13}C$-depleted peak and following $\delta^{13}C$-enriched peak were clearly found at ~125 and ~ 135 cm BL, respectively. The small HSs at ~110 cm BL, shown in **Figure 6**, could be just-weaned animals who might forage independently of their mothers, and be stranded in salmon set nets.

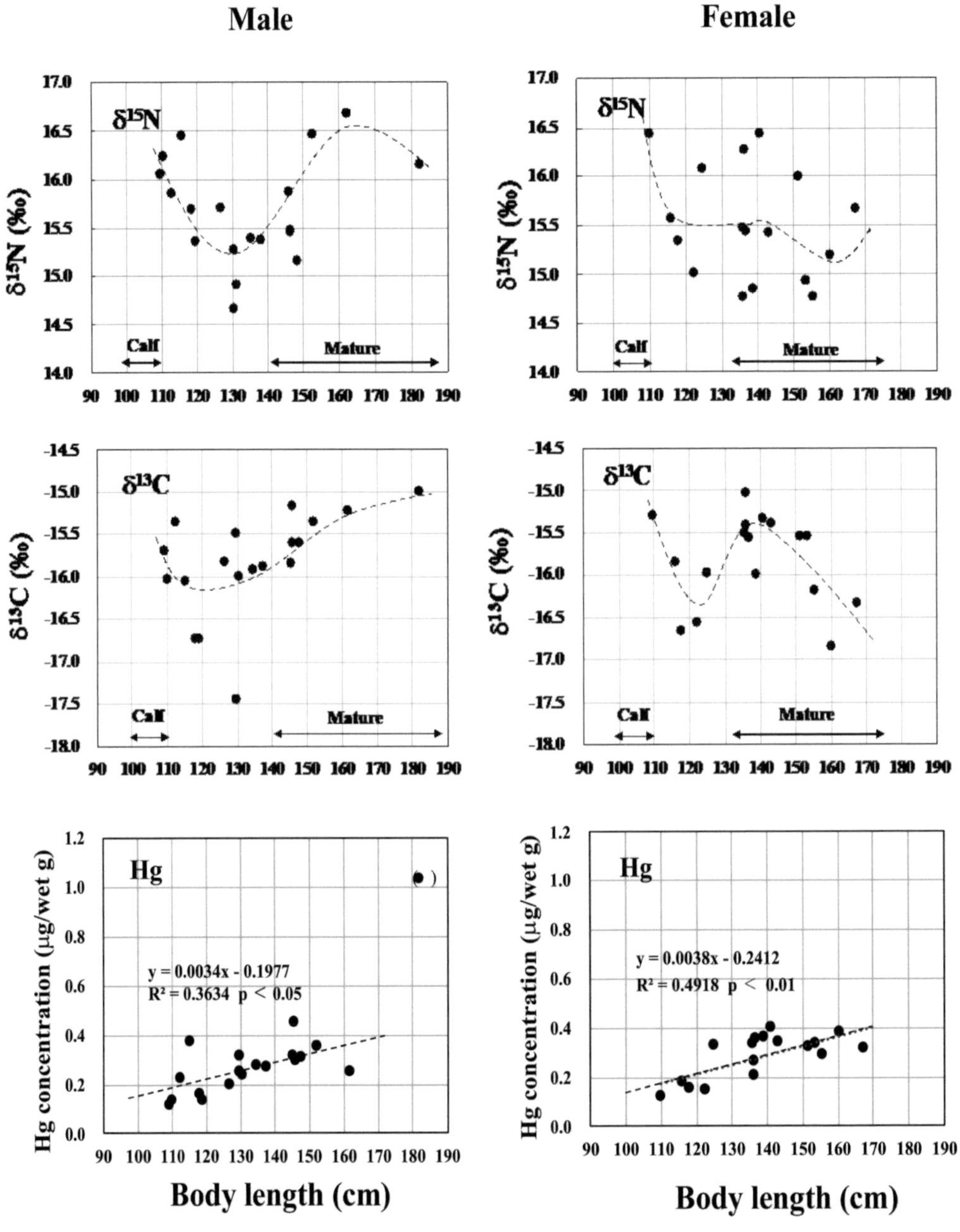

Figure 6.
Ontogenetic changes of $\delta^{15}N$ and $\delta^{13}C$ signatures and Hg concentration in muscle of male and female harbor seals stranded in Hokkaido.

The $\delta^{15}N$-enriched peaks due to the lactation were not observed in HS samples, as our samples did not include nursing pups whose BL was less than 110 cm. Sharp decreases in $\delta^{15}N$ and $\delta^{13}C$ values were found in male and female pups at 110–120 cm BL after weaning, which may reflect their weaning processes: The feeding sifts from milk to solid food and turnover rate of $\delta^{15}N$ signature may be extremely first in HS muscle.

Sex-related differences in $\delta^{15}N$ and $\delta^{13}C$ profiles were observed in the animals. Pregnancy and lactation of females may be the reasons for the differences in $\delta^{15}N$ and $\delta^{13}C$ signatures. Sex-related differences in $\delta^{15}N$ and $\delta^{13}C$ signatures were not found in the MWs (**Figure 4**) and were unclear in DPs (**Figure 5**), which migrate annually in the waters around Japan.

3.5 Isotopic segregation of calves of cetacean species stranded in Hokkaido

Stable isotope analyses of carbon and nitrogen in the tissues of suckling offspring, before weaning, when milk constitutes the entire diet, are increasingly used as proxies for maternal forging habits and trophic levels [54, 60, 79]. This indirect approach is useful when sampling from females is difficult or has a high potential risk. In this section, we compare the $\delta^{13}C$ and $\delta^{15}N$ values of stranded calves (nursing and weaning stages) from mysticete species (MW) and two odontocete species (DP and KW) inhabiting the waters around Hokkaido, Japan, and the North Pacific Ocean, to investigate whether these calves reflect their maternal foraging habits and trophic levels.

Figure 7 shows the $\delta^{13}C$ and $\delta^{15}N$ values of MW, DP, and KW calves stranded along the coast of Hokkaido. MW is opportunistic feeders that temporally and

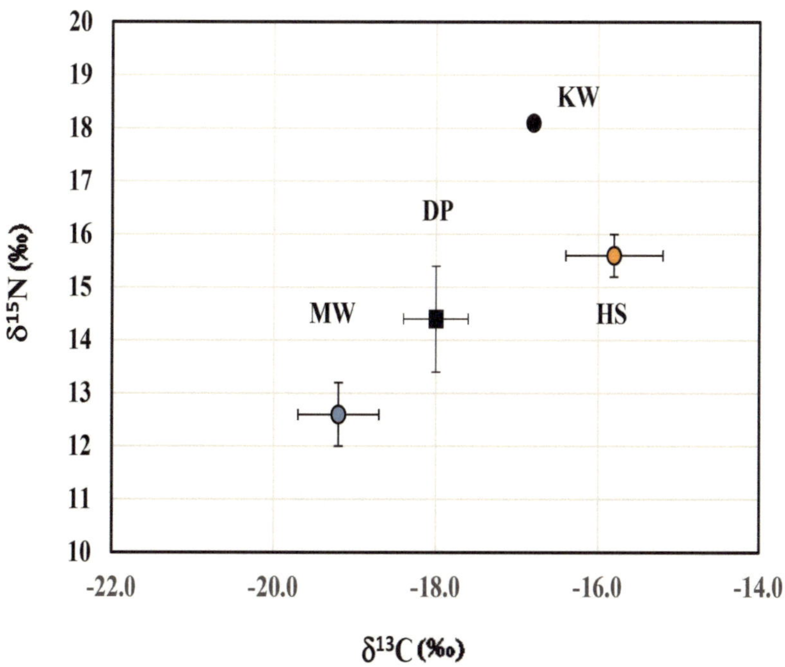

Figure 7.
*Isotopic discrimination of calves of common mink whale, Dall's porpoise and killer whale stranded in Hokkaido. MW: Common minke whales (see **Figure 4**, n = 12), DP: Dall's porpoises (see **Figure 5**, BL ≥130 cm, n = 11), KW: killer whales (see **Table 1**, n = 3), HS: harbor seals (see **Figure 6**, n = 32).*

regionally adapt to prey items (zooplankton and small fish), whereas KW and DP are odontocetes, and the KW is a cetacean located at the top of the marine food chain.

The $\delta^{15}N$ and $\delta^{13}C$ values of KW calves and MW calves were the highest and lowest, respectively, and those values of DP calves were intermediate, reflecting the trophic positions of their mothers.

Stable isotope signatures of HSs (immature and mature animals shown in **Figure 6** (pinniped) were compared with those of MW, DP and KW calves (**Figure** 7). According to the inshore prey comprising cephalopods, crustaceans, and fish, the $\delta^{13}C$ values of HSs were the highest among the marine mammals tested and $\delta^{15}N$ values of HSs were higher than those of DP and MW calves.

4. Conclusion

The $\delta^{15}N$-enriched peaks owing to nursing and weaning were observed in BW, MW, and DP muscle samples, as shown in **Figures 3–5**. The time lag between the dietary shift from milk to solid food and the $\delta^{15}N$ signature in muscle tissue is considered to be small in BWs, and MWs (mysticetes) and probably in DPs (odontocete). Thus, the $\delta^{15}N$ signature in muscle tissues could serve as an excellent proxy for the lactation of calves.

Brief lactation (weaning) could result in a sharp decrease in $\delta^{15}N$ values (**Figures 4** and **6**), whereas prolonged weaning may result in a gradual decrease (**Figure** 5).

The $\delta^{15}N$ and $\delta^{13}C$ values of the muscle of cetacean calves may serve as excellent proxies for maternal forging habits and trophic levels (**Figure** 7).

In marine mammals, the increase in Hg burden in calves could serve as a proxy for the weaning and eating of fish (**Figures 4–6**).

Acknowledgements

We would like to thank Stranding Network Hokkaido (SNH) for providing the cetacean samples and stranding information used in this study. We would like to thank to Inter-Research Science Publisher for permission of reproduce use of Table 5 from Lee et al. (2005) published in Marine Ecology Progress Series. We would like to thank to Aquatic Mammals for granting permission to use Figure 2 from Endo et al. (2021).

Author details

Tetsuya Endo[1]* and Mari Kobayashi[2]

1 School of Pharmaceutical Sciences, Health Sciences University of Hokkaido, Ishikari-Tobetsu, Hokkaido, Japan

2 Faculty of Bioindustry, Tokyo University of Agriculture, Abashiri, Hokkaido, Japan

*Address all correspondence to: endotty531115@gmail.com

References

[1] Dalerum F, Bennett NC, Clutton-Brock TH. Longitudinal differences in ^{15}N between mothers and offspring during and after weaning in a small cooperativemammal, the meerkat (*Suricata suricatta*). Rapid Communications in Mass Spectrometry. 2007;**21**(12):1889-1892. DOI: 10.1002/rcm.3032

[2] Miller JF, Millar JS, Longstaffe FJ. Stable nitrogen and carbon isotope discrimination between juveniles and adults in an income-breeding small mammal (*Peromyscus maniculatus*). Mammalian Biology. 2011;**76**(5):563-569. DOI: 10.1016/j.mambio.2011.02.006

[3] Borrell A, Gómez-Campos E, Aguilar A. Influence of reproduction on stable-isotope ratios: Nitrogen and carbon isotope discrimination between mothers, fetuses, and milk in the fin whale, a capital breeder. Physiology and Biochemical Zoology. 2016;**89**(1):41-50. DOI: 10.1086/684632

[4] Sapriza FGR. Lactation Strategies and Milk Composition in Pinnipeds. Biology, Physiological Basis, Nutritional Requirements, and Modelization. Rijeka: IntechOpen Book Series; 2019. DOI: 10.5772/intechopen.85386

[5] Lockyer C. Review of baleen whale (mysticeti) reproduction and implications for management. Report to International Whaling Commission, Special Issue. 1984;**6**:27-50

[6] Oftedal OT. Lactation in whales and dolphins: Evidence of divergence between baleen- and toothed-species. Journal of Mammary Gland Biology and Neoplasia. 1997;**2**(3):205-230. DOI: 10.1023/A:1026328203526

[7] Oftedal OT. Animal nutrition and metabolism group symposium on 'regulation of maternal reserves and effects on lactation and the nutrition of young animals'. Use of maternal reserves as a lactation strategy in large mammals. Proceedings of the Nutrition Society. 2000;**59**:99-106

[8] Bowen WD. Behavioural ecology of pinniped neonates. In: Renouf D, editor. The Behaviour of Pinnipeds. Berlin: Springer Book Archive; 1991. pp. 66-127. DOI: 10.1007/978-94-011-3100-1_3

[9] Cherel Y, Hobson KA, Guinet C. Milk isotopic values demonstrate that nursing fur seal pups are a full trophic level higher than their mothers. Rapid Communication in Mass Spectrometry. 2015;**29**(16):1485-1490. DOI: 10.1002/rcm.7243

[10] Boecklen WJ, Yarnes CT, Cook BA, James AC. On the use of stable isotopes in trophic ecology. Annual Review of Ecology, Evolution, and Systematics. 2011;**42**:411-440. DOI: 10.1146/annurev-ecolsys-102209-144726

[11] O'Brien DM. Stable isotope ratios as biomarkers of diet for health research. Annual Review of Nutrition. 2015;**35**: 565-594. DOI: 10.1146/annurev-nutr-071714-034511

[12] Hilderbrand GV, Farley SD, Robbins CT, Hanley TA, Titus K, Servheen C. Use of stable isotopes to determine diets of living and extinct bears. Canadian Journal of Zoology. 1996;**74**:2080-2088. DOI: 10.1139/z96-236

[13] Habran S, Debier C, Crocker DE, Houser SD, Lepoint G, Bouquegneau JM, et al. Assessment of gestation, lactation and fasting on stable isotope ratios in northern elephant seals (*Mirounga*

angustirostris). Marine Mammal Science. 2010;**26**(4):880-895. DOI: 10.1111/j.1748-7692.2010.00372.x

[14] Jansen OE, Aarts GM, Das K, Lepoint G, Michel L, Reijnders PJH. Feeding ecology of harbour porpoises: Stable isotope analysis of carbon and nitrogen in muscle and bone. Marine Biology Research. 2012;**8**:829-841. DOI: 10.1080/17451000.2012.692164

[15] Beltran RS, Peterson SH, McHuron EA, Reichmuth C, Hückstädt LA, Costa DP. Seals and sea lions are what they eat, plus what? Determination of trophic discrimination factors for seven pinniped species. Rapid Communication in Mass Spectrometry. 2016;**30**(9):1115-1122. DOI: 10.1002/rcm.7539

[16] Peterson SH, McHuron EA, Kennedy SN, Ackerman JT, Rea LD, Castellini JM, et al. Evaluating hair as a predictor of blood mercury: The influence of ontogenetic phase and life history in pinnipeds. Archives of Environmental Contamination and Toxicology. 2016;**70**(1):28-45. DOI: 10.1007/s00244-015-0174-3

[17] Yurdakök K. Lead, mercury, and cadmium in breast milk. Journal of Pediatric and Neonatal Individualized Medicine. 2015;**4**(2):e040233. DOI: 10.7363/040223

[18] Honda K, Tatsukawa R, Itano K, Miyazaki N, Fujiyama T. Heavy metal concentrations in muscle, liver and kidney tissue of striped dolphin, *Stenella coeruleoalba*, and their variations with body length, weight, age and sex. Agricultural and Biological Chemistry. 1983;**47**(6):1219-1228. DOI: 10.1080/00021369.1983.10863383

[19] Das K, Beans C, Holsbeek L, Mauger G, Berrow SD, Rogan E, et al. Marine mammals from Northeast Atlantic: Relationship between their trophic status as determined by $\delta^{13}C$ and $\delta^{15}N$ measurements and their trace metal concentrations. Marine Environmental Research. 2003;**56**(3):349-365. DOI: 10.1016/S0141-1136(02)00308-2

[20] Dehn LA, Follmann EH, Rosa C, Duffy LK, Thomas DL, Bratton GR, et al. Stable isotope and trace element status of subsistence-hunted bowhead and beluga whales in Alaska and gray whales in Chukotka. Marine Pollution Bulletin. 2006;**52**(3):301-319. DOI: 10.1016/j.marpolbul.2005.09.001

[21] Endo T, Hisamichi Y, Kimura O, Haraguchi K, Baker CS. Contamination levels of mercury and cadmium in melon-headed whales (*Peponocephala electra*) from a mass stranding on the Japanese coast. Science of the Total Environment. 2008;**401**:73-80. DOI: 10.1016/j.scitotenv.2008.04.006

[22] Gerpe MS, de León AP, Bastida R, Moreno VJ, Rodríguez DH. Sharp accumulation of heavy metals after weaning in the South American fur seal *Arctocephalus australis*. Marine Ecology Progress Series. 2009;**375**:239-245. DOI: 10.3354/meps07799

[23] Itano K, Kawai S, Miyazaki N, Tatsukawa R, Fujiyama T. Mercury and selenium levels at the fetal and suckling stages of striped dolphin, *Stenella coeruleoalba*. Agricultural and Biological Chemistry. 1984;**48**(7):1691-1698. DOI: 10.1271/bbb1961.48.1691

[24] Endo T, Haraguchi K, Hotta Y, Hisamichi Y, Lavery S, Dalebout MK, et al. Total mercury, methyl mercury, and selenium levels in the red meat of small cetaceans sold for human consumption in Japan. Environmental Science and Technology. 2005;

39(15):5703-5708. DOI: 10.1021/es050215e

[25] Oskarsson A, Hallén IP, Sundberg J, Grawé KP. Risk assessment in relation to neonatal metal exposure. The Analyst. 1998;**123**(1):19-23. DOI: 10.1039/a705136k

[26] Clarkson TW, Magos L. The toxicology of mercury and its chemical compounds. Critical Reviews in Toxicology. 2006;**36**(8):609-662. DOI: 10.1080/10408440600845619

[27] Castellini JM, Rea LD, Lieske CL, Beckmen KB, Fadely BS, Maniscalco JM, O'Hara TM. Mercury concentrations in hair from neonatal and juvenile Steller Sea lions (*Eumetopias jubatus*): Implications based on age and region in this northern Pacific marine sentinel piscivore. EcoHealth. 2012;**9**(3):267-277. DOI: 10.1007/s10393-012-0784-4

[28] Rea LD, Castellini JM, Correa L, Fadely BS, O'Hara TM. Maternal Steller sea lion diets elevate fetal mercury concentrations in an area of population decline. Science of the Total Environment. 2013;**454–455**: 277-282. DOI: 10.1016/j.scitotenv.2013.02.095

[29] Trukhin AM, Kalinchuk VV. Hair mercury concentrations in the spotted seal (*Phoca largha*) pups from the sea of Japan. Environmental Science and Pollution Research. 2018;**25**(21): 27133-27140. DOI: 10.1007/s11356-018-2731-6

[30] Mann J. Parental behavior. In: Würsing B, Thewissen JGM, Kovacs KM, editors. Encyclopedia of Marine Mammals. 3rd ed. Cambridge, MA: Academic Press. pp. 686-691

[31] Huang SL, Chou LS, Ni IH. Comparable length at weaning in cetaceans. Marine Mammal Science. 2009;**25**(4):875-887. DOI: 10.1111/j.1748-7692.2009.00288.x

[32] Kelly JF. Stable isotopes of carbon and nitrogen in the study of avian and mammalian trophic ecology. Canadian Journal of Zoology. 2000;**78**(1):1-27. DOI: 10.1139/Z99-165

[33] Newsome SD, Clementz MT, Koch PL. Using stable isotope biogeochemistry to study marine mammal ecology. Marine Mammal Science. 2010;**26**(3):509-572. DOI: 10.1111/j.1748-7692.2009.00354.x

[34] Vanderklift MA, Ponsard S. Sources of variation in consumer-diet $\delta^{15}N$ enrichment: A meta-analysis. Oecologia. 2003;**136**:169-182. DOI: 10.1007/s00442-003-1270-z

[35] Caut S, Angulo E, Courchamp F. Variation in discrimination factors ($\Delta^{15}N$ and $\Delta^{13}C$): The effect of diet isotopic values and applications for diet reconstruction. Journal of Applied Ecology. 2009;**46**:443-453. DOI: 10.1111/j.1365-2664.2009.01620.x

[36] Endo T, Hisamichi Y, Kimura O, Haraguchi K, Lavery S, Dalebout ML, et al. Stable isotope ratios of carbon and nitrogen and mercury concentrations in 13 toothed whale species taken from the westerm Pacific Ocean off Japan. Environmental Science & Technology. 2010;**44**:2675-2681. DOI: 10.1021/es903534r

[37] Endo T, Kimura O, Fujii Y, Haraguchi K. Relationship between mercury, organochlorine compounds and stable isotope ratios of carbon and nitrogen in yellowfin tuna (*Thunnus albacares*) taken from different regions of the Pacific and Indian oceans. Ecological Indicators. 2016;**69**:340-347. DOI: 10.1016/j.ecolind.2016.04.021

[38] Valenzuela LO, Sironi M, Rowntree VJ. Interannual variation in the stable isotope differences between mothers and their calves in southern right whales (*Eubalaena australis*). Aquatic Mammals. 2010;**36**(2):138-147. DOI: 10.1578/AM.36.2.2010.138

[39] Drago M, Franco-Trecu V, Cardona L, Inchausti P. Diet-to-female and female-to-pup isotopic discrimination in South American sea lions. Rapid Communication in Mass Spectrometry. 2015;**29**(16):1513-1520. DOI: 10.1002/rcm7249

[40] Endo T, Kimura O, Terasaki M, Kobayashi M. Body length, stable carbon, and nitrogen isotope ratios and mercury levels in common minke whales (*Balaenoptera acutorostrata*)stranded along the coast of Hokkaido, Japan. Aquatic Mammals. 2021;**47**(1):86-95. DOI: 10.1578/AM.47.1.2021.86

[41] Ryan C, McHugh B, Trueman CN, Harrod C, Berrow SD, O'Connor I. Accounting for the effects of lipids in stable isotope (δ^{13}C and δ^{15}N values) analysis of skin and blubber of balaenopterid whales. Rapid Communication in Mass Spectrometry. 2012;**26**(23):2745-2754. DOI: 10.1002/rcm.6394

[42] Polischuk SC, Hobson KA, Ramsay MA. Use of stable-carbon and -nitrogen isotopes to assess weaning and fasting in female polar bears and their cubs. Canadian Journal of Zoology. 2001; **79**:499-511. DOI: 10.1139/cjz-79-3-499

[43] Stegall VK, Farley SD, Rea LD, Pitcher KW, Rye RO, Kester CL, et al. Discrimination of carbon and nitrogen isotopes from milk to serum and vibrissae in Alaska steller sea lions (*Eumetopias jubatus*). Canadian Journal of Zoology. 2008;**86**:17-23. DOI: 10.1139/Z07-115

[44] Borrell A, Abad-Oliva N, Gómez-Campos E, Giménez J, Aguilar A. Discrimination of stable isotopes in fin whale tissues and application to diet assessment in cetaceans. Rapid Communications in Mass Spectrometry. 2012;**26**:1596-1602. DOI: 10.1002/rcm.6267

[45] Roth J, Hobson KA. Stable carbon and nitrogen isotopic fractionation between diet and tissue of captive red fox: Implications for dietary reconstruction. Canadian Journal of Zoology. 2000;**78**(5):848-852. DOI: 10.1139/cjz-78-5-848

[46] Kurle CM, Worthy GAJ. Stable nitrogen and carbon isotope ratios in multiple tissues of the northern fur seal *Callorhinus ursinus*: Implications for dietary and migratory reconstructions. Marine Ecology Progress Series. 2002; **236**:289-300. DOI: 10.3354/meps236289

[47] Miller JF, Millar JS, Longstaffe FJ. Carbon- and nitrogen-isotope tissue–diet discrimination and turnover rates in deer mice, *Peromyscus maniculatus*. Canadian Journal of Zoology. 2008;**86**(7):685-691. DOI: 10.1139/Z08-042

[48] Trueman CN, McGill RAR, Guyard PH. The effect of growth rate on tissue-diet isotopic spacing in rapidly growing animals. An experimental study with Atlantic salmon (*Salmo salar*). Rapid Communication in Mass Spectrometry. 2005;**19**:3239-3247. DOI: 10.1002/rcm.2199

[49] Zanden MJV, Clayton MK, Moody EK, Solomon CT, Weidel BC. Stable isotope turnover and half-life in animal tissues: A literature synthesis. PLoS One. 2015;**10**:e011682. DOI: 10.1371/journal.pone.0116182

[50] Milmann L, de Oliveira LR, Danilevicz IM, Di Beneditto APD,

Botta S, Siciliano S, et al. Stable isotope analysis in baleen whales (suborder: mysticeti): A review until 2017. Boletim Do Laboratório De Hidrobiologia. 2020;**30**(2):1-14. DOI: 10.18764/1981-6421e2020.10

[51] Hobson KA, Schell D, Renouf DY, Noseworthy E. Stable carbon and nitrogen isotopic fractionation between diet and tissues of captive seals. Canadian Journal of Fisheries and Aquatic Sciences. 1996;**53**(3):528-533. DOI: 10.1139/cjfas-53-3-528

[52] Hobson KA, Riget FF, Outridge PM, Dietz R, Born E. Baleen as a biomonitor of mercury content and dietary history of North Atlantic minke whales (*Balaenopetra acutorostrata*): Combining elemental and stable isotope approaches. Science of the Total Environment. 2004; **331**(1–3):69-82. DOI: 10.1016/j.scitotenv. 2004.03.024

[53] Mitani Y, Bando T, Takai N, Sakamoto W. Patterns of stable carbon and nitrogen isotopes in the baleen of common minke whale *Balaenoptera acutorostrata* from the western North Pacific. Fisheries Science. 2006;**72**(1): 69-76. DOI: 10.1111/j.1444-2906.2006. 01118.x

[54] Lowther A, Goldsworthy SD. Detecting alternate foraging ecotypes in Australian sea lion (*Neophoca cinerea*) colonies using stable isotope analysis. Marine Mammal Science. 2011;**27**(3): 567-586. DOI: 10.1111/ j.1748-7692.2010.00425.x

[55] Matthews CJD, Ferguson SH. Weaning age variation in beluga whales (*Delphinapterus leucas*). Journal of Mammalogy. 2015;**96**(2):425-437. DOI: 10.1093/jmammal/gyv046

[56] Aubail A, Dietz R, Rigét F, Simon-Bouhet B, Florence F. An evaluation of teeth of ringed seals (*Phoca hispida*) from Greenland as a matrix to monitor spatial and temporal trends of mercury and stable isotopes. Science of the Total Environment. 2010;**408**(21):5137-5146. DOI: 10.1016/j.scitotenv.2010.07. 038Hobson

[57] Vales DG, Cardona L, Garcia NA, Zenteno L, Crespo EA. Ontogenetic dietary changes in male south American fur seals *Arctocephalus australis* in Patagonia. Marine Ecology Progress Series. 2015;**525**:245-260. DOI: 10.3354/ meps11214

[58] Endo T, Kimura O, Sato R, Kobayashi M, Matsuda S, Matsuishi T, et al. Stable isotope ratios of carbon, nitrogen and oxygen in killer whales (*Orcinus orca*) stranded on the coast of Hokkaido, Japan. Marine Pollution Bulletin. 2014;**86**(1–2):238-243. DOI: 10.1016/j.marpolbul.2014.07.012

[59] Endo T, Kimura O, Hisamichi Y, Minoshima Y, Haraguchi K. Age-dependent accumulation of heavy metals in a pod of killer whales (*Orcinus orca*) stranded in the northern area of Japan. Chemosphere. 2007;**67**:51-59. DOI: 10.1016/j.chemosphere.2006.09.086

[60] Ducatez S, Dalloyau S, Richard P, Guinet C, Cherel Y. Stable isotopes document winter trophic ecology and maternal investment of adult female southern elephant seals (*Mirounga leonina*) breeding at the Kerguelen Islands. Marine Biology. 2008;**155**(4): 413-420. DOI: 10.1007/s00227-008-1039-3

[61] Jenkins SG, Partridge ST, Stephenson TR, Farley SD, Robbins CT. Nitrogen and carbon isotope fractionation between mothers, neonates, and nursing offspring. Oecologia. 2001;**129**(3):336-341. DOI: 10.1007/s004420100755

[62] Haraguchi K, Hisamichi Y, Endo T. Accumulation and mother-to-calf transfer of anthropogenic and natural organohalogens in killer whales (*Orcinus orca*) stranded on the Pacific coast of Japan. Science of the Total Environment. 2009;**407**: 2853-2859. DOI: 10.1016/j.scitotenv.2009.01.003

[63] Doi H, Akamatsu F, Gonzále AJ. Starvation effects on nitrogen and carbon stable isotopes of animals: An insight from meta-analysis of fasting experiments. Royal Society of Open Science. 2017;**4**:170633. DOI: 10.1098/rsos.170633

[64] Chilvers BL. Isotope values from milk and blood serum in New Zealand sea lions: Are pups feeding on milk a trophic level higher than their mothers? Marine Biology. 2021; **168**(1):12. DOI: 10.1007/s00227-020-03817-4

[65] Haraguchi K, Endo T, Sakata M, Masuda Y, Simmonds MS. Contamination survey of heavy metals and organochlorine compounds in cetacean products purchased from Japan. Food Hygiene and Safety Science. 2000; **41**(4):287-296. DOI: 10.3358/shokueishi.41.287

[66] Simmonds MP, Haraguchi K, Endo T, Cipriano F, Palumbi SR, Troisi GM. Human health significance of organochlorine and mercury contaminants in Japanese whale meat. Journal of Toxicology and Environmental Health Part A. 2002; **65**(17):1211-1225. DOI: 10.1080/152873902760125714

[67] Endo T, Hotta Y, Haraguchi K, Sakata M. Mercury contamination in the red meat of whales and dolphins marketed for human in Japan. Environmental Science & Technology. 2003;**37**:2681-2685. DOI: 10.1021/es034055n

[68] Bustamante P, Caurant F, Fowler SW, Miramand P. Cephalopods as a vector for the transfer of cadmium to top marine predators in the north-east Atlantic Ocean. Science of Science of the Total Environment. 1998; **20**(1):71-80. DOI: 10.1016/S0048-9697(98)00250-2

[69] Yasunaga G, Fujise Y. Temporal trends and factors affecting mercury levels in common minke, Bryde's and sei whales and their prey species in the western North Pacific. Reports of International Whaling Commission. 2009; SC/J09/JR23. Available from: www.icrwhale.org/pdf/SC-J09-JR23.pdf

[70] Endo T, Hayasaka M, Hisamichi Y, Kimura O, Haraguchi K. Carbon and nitrogen stable isotope ratios and mercury concentration in the scalp hair of resident from Taiji, a whaling town. Marine Pollution Bulletin. 2013;**69**(1–2): 116-121. DOI: 10.1016/j.marpolbul.2013.01.018

[71] Brookens TJ, O'Hara TM, Taylor RJ, Bratton GR, Harvey JT. Total mercury body burden in Pacific harbor seal, *Phoca vitulina richardii*, pups from Central California. Marine Pollution Bulletin. 2008;**56**(1):27-41. DOI: 10.1016/j.marpolbul.2007.08.010

[72] Lee SH, Schell DM, McDonald TL, Richardson WJ. Regional and seasonal feeding by bowhead whales *Balaena Cetus* as indicated by stable isotope ratios. Marine Ecology Progress Series. 2005;**285**:271-287. DOI: 10.3354/MEPS285271

[73] Horstmann-Dehn L, Follmann EH, Rosa C, Zelensky G, George C. Stable carbon and nitrogen isotope ratios in muscle and epidermis of arctic whales.

Marine Mammal Science. 2012;**28**(2): E173-E190. DOI: 10.1111/j.1748-7692. 2011.00503.x

[74] George JC, Bada J, Zeh J, Scott L, Brown SE, O'Hara T, et al. Age and growth estimates of bowhead whales (*Balaena mysticetus*) via aspartic acid racemization. Canadian Journal of Zoology. 1999;**77**:571-581. DOI: 10.1139/z99-015

[75] Nakamura T, Kimura O, Matsuda A, Matsuishi T, Kobayashi M, Endo T. Radiocesium contamination of cetaceans stranded along the coast of Hokkaido, Japan, and an estimation of their travel routes. Marine Ecology Progress Series. 2015;**535**:1-9. DOI: 10.3354/meps11432

[76] Kato H. Body length, reproduction and stock separation of minke whales off northern Japan. Report to International Whaling Commission. 1992;**42**:443-453

[77] Kasuya T. The life history of Dall's porpoise with special reference to the stock off the Pacific coast of Japan. The Scientific Reports of the Whales Research Institute. 1978;**30**:1-63

[78] Naito Y, Nishiwaki M. The growth of two species of the habour seal in the adjacent waters of Hokkaido. The Scientific Reports of the Whales Research Institute. 1972;**24**:127-144

[79] Aurioles D, Koch PL, Le Boeuf BJ. Differences in foraging location of Mexican and California elephant seals: Evidence from stable isotopes in pups. Marine Mammal Science. 2006;**22**(2): 326-338. DOI: 10.1111/j.1748-7692.2006.00023.x

Chapter 5

Perspective Chapter: Status of Dolphin in the Maritime Area of Bangladesh

Md. Muzammel Hossain

Abstract

The aquatic mammalian species is the best indicator for the health of water specially dolphins. Due to human anthropogenic activities, dolphin survival rate and movement are reduced. The dreadful conditions of coastal habitats can have major concerns for dolphin population and distribution. Some dolphins swim in a short distance and some swim in a long distance. Southeast Asia is a significant geographical region for dolphin conservation. Several dolphins are highly important for the maritime area of Bangladesh that were sighted in the coastal and marine water such as Irrawaddy dolphin, Indo-Pacific Humpback dolphin, Bottlenose dolphin, Spotted dolphin, Spinner dolphin, and Ganges dolphins. Marine protected area (MPA) is a valuable zone for dolphin conservation as well as biological species. This primary information of dolphins helps for further investigation in the Bangladeshi water. The research action plan must be considered with coastal habitat, marine protected area and fishing community to conserve dolphins. We should be concerned about dolphin conservation through local and international community to develop the environment and the blue economy. Local community directly involved in the maritime area due to livelihood opportunities.

Keywords: aquatic mammal, dolphin, habitat, conservation

1. Introduction

Among the cetacean, the dolphin is the common name of aquatic mammals. Over the last decades, aquatic mammal conservation is a global concern due to human anthropogenic activities and rapidly declining the biodiversity. Losses of aquatic mammals or biodiversity directly affect on ecosystem locally or internationally that concerned by the Convention of Biological Diversity (CBD) in 1992. Marine mammal conservation is continuing process which cannot ever be well-thought-out completely in south Asia. In the Bay of Bengal, Swatch of No-ground has been established as Marine Protected Area (MPA) for biological species. As a big delta basin Ganges, Brahmaputra, and Meghna river systems are present along with the coastal regions in Bangladesh. IUCN Bangladesh [1] reported 10 cetacean species are present in the aquatic ecosystem of Bangladesh include the Ganges river dolphin. All marine dolphins are included in the Family Delphinidae. The very little study occurred on the dolphins in the maritime area of Bangladesh, except some reports of species events [2–8]. According to WCS [9] about 1738 km^2 areas are considered as a marine protected area (MPA) for marine mammal species, and MPA is constructed at the Northwest and Southeast geographic area of Bangladesh that is recognized with the Indian Ocean in **Figure 1**.

However, some author little studies completed on dolphins [7–12] but the status of dolphin investigation is not sufficient in the maritime area of Bangladesh due to lack of funds, strategic plan, and awareness. Although, False Killer whale, Bryde's whale, and Sperm whales are found in the marine water but different types of dolphins are survived in maritime areas of Bangladesh.

2. Dolphins in the maritime area

The species conservation effort should be a concern with ecological, behavioral, population, and species diversity. Among the mammal Irrawaddy dolphin, Indo-Pacific Humpback dolphin, Bottlenose dolphin, Spotted dolphin, Spinner dolphin, and Rough-toothed dolphin are present in the maritime area of Bangladesh in **Figure 2**. Only Ganges river dolphins survive in the freshwater, coastal water, and the mouth of the Ganges at Sundarban area.

2.1 Irrawaddy dolphin

This dolphin name comes from the Irrawaddy River, Myanmar and endangered species among all cetacean species in the world. Its scientific name is *Orcaella brevirostris* and its local name is the Iraboti dolphin. This dolphin is considered Vulnerable globally and Near Threatened in Bangladesh due to anthropogenic activities. Irrawaddy dolphins are critically endangered. Species color is gray to dark slate blue, paler underneath, rounded, and small dorsal fin present; the flipper is rounded and wide. About 451 individual Irrawaddy dolphins were sighted near the coastal water of Sundarbans mangrove forest [12]. This dolphin is very similar to the finless porpoise and fishes, crustaceans, and invertebrates are main food [13]. In the coastal region, they are more limited although survive around South and Southeast Asia. According to Smith et al. [14] the 779.7 km of trackline and a 16,779 km^2 study

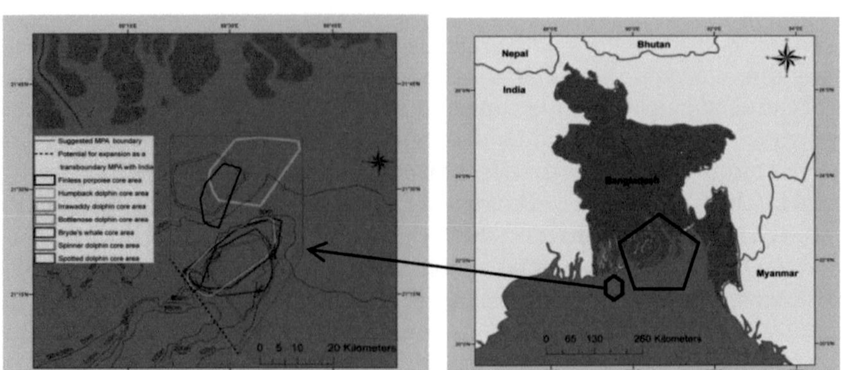

Figure 1.
Map of dolphin occurrence in the maritime area, black color line box indicated site (WCS, 2015).

Figure 2.
Different types of dolphin occurred in the maritime area of Bangladesh. (1) Irrawaddy dolphin, (2) Indo-Pacific Humpback dolphin, (3) Indo-Pacific Bottlenose dolphin, (4) Spotted dolphin, (5) Spinner dolphin, and (6) Ganges river dolphin.

area were investigated, an abundance of Irrawaddy dolphin was assessed 5383 (CV = 39.5%) in the maritime water in Bangladesh including number (n) was 75, mean group size was 2.2, SD 1.8, dolphin range was 1–10. Also, large populations of Irrawaddy dolphins were recorded in the marine coastal water of Bangladesh. The reported number of Irrawaddy dolphins 114 from 2004 to 2014 [15] and UNDP Bangladesh [16] also stated a total number of 451, respectively in the Sundarban mangrove region of Bangladesh in **Figure 3**. Smith et al. [12] reported the encounter rate of dolphins as 0.19 in the Sundarban area, Bangladesh whereas a similar encounter of 0.16 showed at the Ayeyarwady River in Myanmar [17]. The Sundarbans and their adjacent areas are major and hotspot areas for the Irrawaddy Dolphin and half of the population are present in the maritime area of Bangladesh. Based on IUCN Red List in Bangladesh and India, this species is Vulnerable, but Critically Endangered species in Laos, Myanmar, Malaysia, Thailand, and Philippines.

2.2 Indo-Pacific Humpback dolphin

The Indo-Pacific humpback dolphin has been recorded in the marine water of Bangladesh and also sighted near the eastern India region [18]. This species scientific name is *Sousa chinensis* and its local name is Golapi dolphin. The highest size recorded

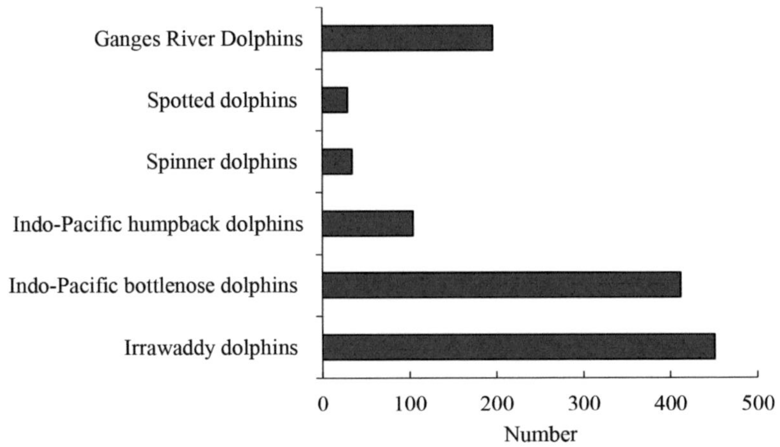

Figure 3.
Number of species sighted in the maritime water of Bangladesh.

280 cm, healthy body, and extended beak. Color outline differs with age and area. Color show bluish gray to light cream or pink. This dolphin shape is very similar with the bottlenose dolphin. Color, dorsal fin, and head size separate them. Humpback dolphin usually swims in groups. They are survived in the tropics and subtropics area of coastal water and small fishes are major food. The marine water environment of Southeast Asia is a significant habitat for this dolphin but this species population, distribution, abundance, and feature of habitat very little study has been done to ensure survival species [19–21]. World Conservation Society [9] remarks the area about 327 km^2 of importance habitat for humpback dolphins encompassing 73% of sightings in coastal waters near the mangrove forest. Also, at the 24.73°–24.33°N geographic site, this dolphin sighting per unit effort was 3.88 [22, 23]. WCS was assessed humpback dolphins survive in shallow water and coastal waters in the Bay of Bengal but most of the species occur in the northern Indian and eastern Pacific oceans. According to the IUCN Red List this species is Near Threatened but presently like remark as vulnerable. Most of the humpback dolphins swim in the marine protected area (MPA).

2.3 Indo-Pacific Bottlenose dolphin

This mammal scientific name is *Tursiops aduncus* and its local name is the Botolnak dolphin. The highest size recorded 250 cm; also have a snout, large flipper, dorsal fin, and flukes. The color outline is gray and marginally dark present. Although, about 402 Indo-Pacific bottlenose dolphins are sighted from 2004 to 2014 and also large population was recorded in the marine protected area in **Figure 3**. WCS [9] reported about 1700–2200 individual Indo-Pacific bottlenose dolphins that survived in the Bangladesh area. This species is also sighted in the face of the Swatch-of-No-Ground which is a highly biological habitat in the Bay of Bengal. Although, IUCN considers the species as data deficient (DD) but due to their distribution and population remark as vulnerable in the environment of the Bay of Bengal. Fisherman occasionally caught this dolphin by gillnets and seines net during fishing. WCS considers about 282 km^2 as Marine Protected Area (MPA) for this dolphin whereas 90% of species are sighted.

2.4 Spotted dolphin

The spotted dolphin is the most dominant animal in the cetacean groups in tropical area. Generally, Pantropical spotted dolphins occurred in the marine protected areas of the Bay of Bengal. This dolphin's scientific name is *Stenella attenuate* and its local name is Chitra dolphin. Its body structure is slender and looks like streamlined. The dorsal fin is narrow and pointed at the tip. The spot is present in adult dolphins. The abdomen color is a brighter gray. Maximum sizes of spotted dolphins found about 260 cm long. WCS reported 29 sightings of spotted dolphins and an average group size of 84 in the marine protected area (MPA) [9]. The spotted dolphin population is reducing due to unplanned fishing in the Bangladesh region. IUCN is considered as least concern and very little study occurred in Bangladesh on this dolphin. In the Swatch-of-No-Ground, about 86% of spotted dolphins occurred due to well biological habitat and 263 km^2 areas considered as MPA for this species.

2.5 Spinner dolphin

The Spinner dolphin's scientific name is *Stenella longirostris* and the local name is Ghurni dolphin. This dolphin's body shape is slender with a particularly long beak. Adult spinner dolphin's maximum size recorded is 240 cm. Generally male is larger than female. Triangular to curved shape present in the dorsal fin. Three types of color are present in Spinner dolphin. The dorsal cape is dark, body sides are light gray, and belly is white. This dolphin is show leaping and spinning activities. Usually, this dolphin survives in the tropical and subtropical waters at geographical areas 40°N and 40°S. Large group of populations are swim in the marine protected area. Some fisheries reported a yearly mortality rate of about 100–1000 dolphins in the marine water due to fishing interaction. Although, IUCN was considered as data deficient but WCS [9] reported 34 sightings and an average group size of 97 in the Swatch-of-No-Ground and 550 individuals was present. Sighting rate 91% in the marine protected area and 263 km^2 areas consider for conservation.

2.6 Ganges river dolphin

This dolphin scientific name is *Platanista gangetica gangetica* (Roxburgh, 1801) and local people call "shusuk" as local name in Bangla. A total of 34 sightings of dolphin, mean density was 0.38 (SD ± 0.37) dolphins/km^2 and best-high-low estimated of 10–12–8 in November 2012 in the Buriganga River area [7]. Also, 62 sightings of dolphins and best-high-low estimated of 9–11–7 in august, monitored by October with 8–10–7 individuals in the Turag River [8]. Another one reported encounter rate of Gangs dolphin was 0.47 in Sundarbans area in **Table 1**. Dolphin group size estimated by the Lincoln-Petersen model and Huggins conditional likelihood model which abundance showed 196 individuals (CV = 12.7%) and 225 individuals (CV = 12.6%), respectively in mangrove channels of the Sundarbans area [12]. It is a freshwater dolphin species scattered throughout the Ganges, Brahmaputra, and Meghna river systems along with the coast area of Bangladesh, Nepal, India, and conceivably Bhutan and 266 dolphins encountered in the entire Brahmaputra river system [11, 26, 27]. In the Indian subcontinent, Ganges dolphin is a special aquatic mammal in the riverine ecosystem. The survey was conducted for Ganges dolphins especially in Turag River and Buriganga River that is a part of the Ganga basin. Department of Environment, Government of Bangladesh Turag and Buriganga River area consider

Geographic area	Encounter rates	Reference(s)
Sundarbans area	0.47	[12]
Sangu River, Bangladesh	1.36	[11]
Karnaphuli River, Bangladesh	0.47	[11]
Buriganga River, Bangladesh	0.38	[7]
Chambal River, India	0.27	[24]
Bhagirathi River, India	0.37	[25]

Table 1.
Encounter rates of Ganges river dolphin.

as "Ecologically Critical Area (ECA)" [28, 29]. This species is enlisted in the CITES and Bangladesh Wildlife (Conservation & Security) Act, 2012, and it is categorized as an endangered species due to the many anthropogenic activities and natural hazards included, such as, water pollution, traffic boat and ship, brickfield, industrial waste, water level reduce.

3. Threats and conservation

Mortality of dolphins in fishing nets and the highest level of water pollution through oil, plastic, and industrial waste are the extreme threats. Illegal fishing such as poisoning or electro fishing and large ship boat, vessel strikes, overfishing, sand and resources mining are also threat for dolphins. In Bangladeshi water, most of the dolphin species death occurs by fishing net specially gill net. Dolphin killing and hunting was extensive in several countries that also threat for the environment [2, 30, 31]. UNDP Bangladesh [16] reported that 130 dolphins were killed from 2007 to 2016 in the waters of Bangladesh which most of the species' death by fishing nets and propellers of ships. GoB [32] also reported a total of 52 dolphins were lost by gillnets in the Sundarban area. Dolphin killing occurred due to collecting body parts has required in fisheries that are used as dolphin oil [33]. Although, several organizations are trying to conserve dolphins but it's not enough, such as SharkLab, United Nations Development Program (UNDP), the Global Environment Facility (GEF), the Forest Department, IUCN Bangladesh, and World Conservation Society (WCS). Dolphin conservation is highly important for other aquatic species in the ecosystem as well as the better health regulator of water. Bangladesh Government has been attentive for the marine protected area (MPA) to conserve cetaceans whereas most of the dolphin sightings occurred in the water. We should concern about bycatch through fishing gear, awareness, and biological data monitoring to conserve dolphins as well as their habitat. Six sanctuaries have been recognized as conservation areas by the Government of Bangladesh for marine mammal and other species. The Swatch of no-Ground is also known as a special conservation area for biological resources.

4. Conclusions

The present condition does not well promise for the future plan of the dolphins. The different key programs should consider improving dolphin conservation. Such as taxonomy study through morphologically and genetically, impact and risk assessment, habitat management assessment, ocean traffic plan, population size and movement, ecology, water quality, environmental hazards and threat, etc. Also season wise fisheries interaction should high attention because dolphin mortality rate increases due to fishing and human anthropogenic activities. Monitoring programs of dolphins, their primary habitat, size, structure, and population can guide or lead delta plan, conservation plan, and climate movement.

Acknowledgements

We would like to express thank all members who contributed to the dolphin research. Also, thanks to Dr. Mohammad Saiful Islam, chairman, Department of Zoology, JnU for his special discussion on marine mammals. In specific, we would like to express our heartfelt gratitude to Rafi, Nur Alam, Kisor Kumar Sarkar, and Hasan Al Razi. Special thanks to Ifat Jahan for picture editing and collection. We also thank an unsigned assessor for useful notes on a former draft.

Author details

Md. Muzammel Hossain[1,2]

1 Department of Zoology, Jagannath University, Dhaka, Bangladesh

2 Biodiversity Conservation and Fisheries Research Center, Dhaka, Bangladesh

*Address all correspondence to: muzammel3@gmail.com

References

[1] IUCN Bangladesh. Red List of Bangladesh Volume 2: Mammals. Bangladesh Country Office, Bangladesh, Dhaka: International Union for Conservation of Nature (IUCN); 2015

[2] Anderson J. Anatomical and Zoological Researches: Comprising an Account of Zoological Results of the Two Expeditions to Western Yunnan in 1868 and 1875; and a Monograph of the Two Cetacean Genera, *Platanista* and *Orcella* [sic]. London: Bernard Quaritch; 1879

[3] Morzer Bruyns WFJ. Some notes on the Irrawaddy dolphin, *Orcaella brevirostris* (Owen, 1866). Mammalian Biology (formerly Zeitschrift für Säugetierkunde). 1966;**31**:367-370

[4] Kasuya T, Haque AKMA. Some informations [sic] on the distribution and seasonal movement of the Ganges dolphin. Scientific Reports of the Whales Research Institute (Tokyo). 1972;**24**:109-115

[5] Khan MAR. Wildlife of Bangladesh: A Checklist. Dhaka, Bangladesh: City Press; 1982

[6] Sarkar SU, Sarkar NJ. Wildlife of Bangladesh: A Systematic List with Status, Distribution and Habitat. Dhaka, Bangladesh: The Rico Printers; 1988

[7] Hossain MM, Alam SMI, Baki MA, Bhouiyan NA. Ganges river dolphin. The Bulletin of the Ecological Society of America. 2016;**97**(2):183-187

[8] Baki MA, Bhouiyan NA, Alam SMI, Islam MS, Hossain MM. Present status of Ganges River dolphins *Platanista gangetica gangetica* (Roxburgh, 1801) in the Turag River, Dhaka, Bangladesh. International Journal of Zoology. 2017;**2017**:1-7. DOI: 10.1155/2017/8964821

[9] World Conservation Society (WCS). Proposal to establish a marine protected area in the Swatch-of-No-Ground submarine canyon and surrounding coastal waters in the Bay of Bengal. 2015. pp. 1-13. Available from: https://www.cbd.int/doc/meetings/mar/ebsaws-2015-01/other/ebsaws-2015-01-gobi-submission6-en.pdf

[10] Alam SMI, Hossain MM, Baki MA, Bhouiyan NA. Status of Ganges dolphin, *Platanista gangetica gangetica* (Roxburgh, 1801) in the river Buriganga, Dhaka. Bangladesh Journal of Zoology. 2015;**43**(1):109-120. DOI: 10.3329/bjz.v43i1.26143

[11] Smith BD, Ahmed B, Ali ME, Braulik G. Status of the Ganges river dolphin or shushuk *Platanista gangetica* in Kaptai Lake and the southern rivers of Bangladesh. Oryx. 2001;**35**:61-72

[12] Smith BD, Braulik G, Strindberg S, Ahmed B, Mansur R. Abundance of Irrawaddy dolphins (*Orcaella brevirostris*) and Ganges river dolphin (*Platanista gangetica gangetica*) estimated using concurrent counts made by independent teams in waterways of the Sundarbans mangrove forest of Bangladesh. Marine Mammal Science. 2006;**22**:527-547

[13] Grill P. The Little Guide: Whales, Dolphins and Porpoises. San Francisco: Fog City Press; 2000

[14] Smith BD, Ahmed B, Mowgli RM, Strindberg S. Species occurrence and distributional ecology of nearshore cetaceans in the Bay of Bengal, Bangladesh, with abundance estimates for Irrawaddy dolphins *Orcaella*

brevirostris and finless porpoises *Neophocaena phocaenoides*. Journal of Cetacean Research and Management. 2008;**10**(1):45-58

[15] Smith BD. Irrawaddy dolphin, *Orcaella brevirostris*. In: Würsig B, Thewissen JGM, Kovacs KM, editors. Encyclopedia of Marine Mammals. San Diego: Elsevier; 2017

[16] UNDP Bangladesh. Community people to become Sundarbans dolphin saviours. 6 January 2019. Available from: https://www.bd.undp.org/content/bangladesh/en/home/presscenter/articles/2019/01/04/Community_Sundarbans_dolphin_saviours.html

[17] Smith BD, Hobbs L. Status of Irrawaddy dolphins *Orcaella brevirostris* in the upper reaches of the Ayeyarwady River, Myanmar. The Raffles Bulletin of Zoology. 2002;**10**(Suppl):67-74

[18] Jefferson TA, Smith BD. Re-assessment of the conservation status of the Indo-Pacific Humpback dolphin (*Sousa chinensis*) using the IUCN red list criteria. Advances in Marine Biology. 2016;**73**:1-26. DOI: 10.1016/bs.amb.2015.04.002

[19] Griffen BD, Drake JM. Effects of habitat quality and size on extinction in experimental populations. Proceedings of the Royal Society B: Biological Sciences. 2008;**275**:2251-2256

[20] Huang SL, Hao Y, Mei Z, Turvey ST, Wang D. Common pattern of population decline for freshwater cetacean species in deteriorating habitats. Freshwater Biology. 2012;**57**:1266-1276

[21] Huang SL, Chang WL, Karczmarski L. Population trend and vulnerability of humpback dolphins *Sousa chinensis* off the west coast of Taiwan. Endangered Species Research. 2014;**26**:147-159

[22] Chou LS. The Population Estimation and Habitats of Cetacean in Coastal Waters of Taiwan (I) (in Chinese). Taipei: Fishery Agency, Council of Agriculture, Executive Yuan; 2007. p. 63

[23] Chou LS, Lee JD. Habitat Hotspot of Humpback Dolphin, *Sousa chinensis*, and Master Planning for Conservation Management (in Chinese). Taipei: Forestry Bureau, Council of Agriculture, Executive Yuan; 2010. p. 65. Available from: http://conservation.forest.gov.tw/public/Attachment/08512385771.pdf

[24] Sharma RK, Mathur R, Sharma SS. Status and distribution of fauna in National Chambal Sanctuary, Madhya Pradesh. The Indian Forester. 1995;**121**:912-916

[25] Sinha RK. Status and conservation of Ganges River dolphin in Bhagirathi Hooghly River systems in India. International Journal of Ecology and Environmental Sciences. 1997;**23**:343-355

[26] Mohan RSL, Dey SC, Bairagi SP, Roy S. On a survey of the Ganges River dolphin *Platanista gangetica* of Brahmaputra River, Assam. Journal of the Bombay Natural History Society. 1997;**94**:483-495

[27] Sinha RK, Smith BD, Sharma G, Prasad K, Choudhury BC, Sapkota K, et al. Status and distribution of the Ganges susu (*Platanista gangetica*) in the Ganges river system of India and Nepal. In: Reeves RR, Smith BD, Kasuya T, editors. Biology and Conservation of Freshwater Cetaceans in Asia. Gland, Switzerland: Occasional Paper of the IUCN Species Survival Commission No. 23; 2000. pp. 42-48

[28] DoE (Department of Environment). The General Over View of Pollution Status of River of Bangladesh. Dhaka, Bangladesh: Department of

Environment, Ministry of Environment and Forest, Government of the People's Republic of Bangladesh; 2001

[29] Mobin MN, Islam MS, Mia MY, Bakali B. Analysis of physicochemical properties of the Turag River water, Tongi, Gazipur in Bangladesh. Journal of Environmental Science and Natural Resources. 2014;7(1):27-33

[30] McNair JFA. Oral Tradition from the Indus. Brighton: R. Gosden; 1908

[31] Pilleri G. Field observations carried out on the Indus dolphin *Platanista indi* in the winter of 1972. Investigations on Cetacea. 1972;**4**:23-29

[32] GoB. Community-Based Resources Management Plan (CBRMP) of the Wildlife Sanctuaries for Dolphins in Bangladesh Sundarbans. Dhaka: GoB; 2018

[33] Bairagi SP. Oil bait fishery of catfishes in Brahmaputra River affecting river dolphin populations in Assam, India. Journal of the Bombay Natural History Society. 1999;**96**:424-426

Chapter 6

Marine Mammals in Syria

Adib Saad and Ilene Mahfoud

Abstract

The Syrian marine water is one of the least studied areas for cetaceans in the Mediterranean Sea. Lack of basic knowledge, such as species composition and habitat, makes it impossible to develop effective conservation measures. The survey carried out along the Syrian coast by monitoring the stranding individuals on the shore since 2002 showed that there were 11 species of marine mammals living in/or visiting the Syrian marine waters at present of which 10 species belonging to the cetacean order and on belonging to pinnipeds order. The following species have been recorded: *Pseudorca crassidens*, *Megaptera novaeangliae*, *Physeter macrocephalus*, *Tursiops truncatus*, *Stenella coeruleoalba*, *Delphinus delphis*, *Ziphius cavirostris*, *Grampus griseus*, *Balaenoptera physalus*, *Balaenoptera acutorostrata*, and *Manchus manchus*. On the other hand, there are four species whose presence in the Syrian marine waters was mentioned a century ago by the researcher Gruvel and his team during three missions (1929–1931), but neither alive nor dead have been seen in this area during the surveys that were carried out since 1996 until the present, these species are: *Phocoena phocoena*, *Globicephalus melas*, *Phocoena communis*, *Hyperoodon rostratus*, *Balaenoptera musculus*. These observations reflect the vulnerability of marine mammals to anthropogenic activities, such as fishing operations, shipping, seismic activities, and climate change.

Keywords: marine mammals, cetaceans, stranding, climatic changes, Mediterranean Sea, Syria

1. Introduction

Although the Mediterranean constitutes only less than 1% of the total area of the world's oceans and seas, it contains 18% of the world's marine biodiversity and contains 28 species of marine mammals (resident, visiting, or roving) that have occurred in the Mediterranean. The species that studies and surveys have proven to be endemic to the Mediterranean are: 11 species of cetaceans: fin whale (*Balaenoptera physalus*); sperm whale (*Physeter macrocephalus*); Cuvier's beaked whale (*Ziphius cavirostris*); short-beaked common dolphin (*Delphinus delphis*); long-finned pilot whale (*Globicephala melas*); Risso's dolphin (*Grampus griseus*); killer whale (*Orcinus orca*); striped dolphin (*Stenella coeruleoalba*) rough-toothed dolphin (*Steno bredanensis*); common bottlenose dolphin (*Tursiops truncatus*); harbor porpoise (*Phocoena phocoena relicta*) and the Mediterranean monk seal (*Monachus monachus*) have adapted well to the region's environmental conditions, but their coexistence with humans is problematic. All the regular species are represented in the Mediterranean by populations genetically distinct from their North Atlantic relatives. Seventeen other species (3 fins and 14 cetaceans) are considered non-endemic, but rather visitors to the Mediterranean heat coming from adjacent regions for various reasons [1]. The Mediterranean is experiencing severe changes as a result of high levels of human activity pressure and its synergistic interaction with the effects of climate change which has affected marine biodiversity [2].

Interactions between marine mammals and commercial fisheries have occurred for centuries and the interactions do not seem to decline. Bycatch, i.e., the unwanted or incidental catch of species other than the target species, is a severe problem in conservation biology and a potential threat to the future survival of marine mammal populations. Marine mammal's populations decline is considered to be an important issue in terms of the biodiversity of vulnerable species and numerous cases of stranding have been documented in the Mediterranean Sea were particularly in the eastern part of the basin [3].

This chapter includes: (1) A review of the available information and data about marine mammals in Syria a century ago; (2) Presentation of the results of the continuous monitoring that we started implementing in 2002 on the stranding of individuals of marine mammals on the Syrian coast; (3) Results of a field survey study at sea onboard research ships Yunis S for a week during August 2008 in partnership with a team from the Faculty of Fisheries at Istanbul University, within the framework of a joint Syrian-Turkish-Lebanese research program; (4) Results of a field survey study at sea that was carried out onboard an Okeanos boat during the month of September 2019; (5) Presentation of the threats that threaten marine mammals in Syria; (6) Presentation of laws and regulations related to the protection of marine mammals in Syria.

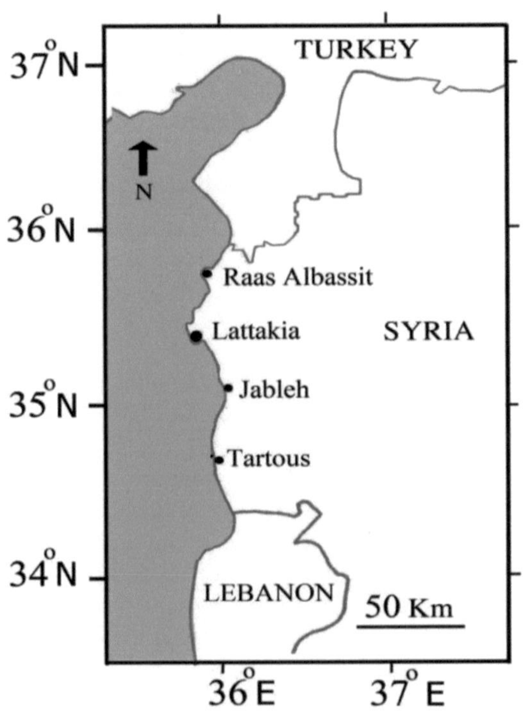

Figure 1.
Map of the Syrian coast on which the most important cities and ports are located, near which the strandings of cetaceans and seals have been recorded during the last 20 years.

We have included below the map of the Syrian coast (**Figure 1**) to clarify the approximate geographical locations that are mentioned when talking about the stranding sites of cetaceans and monk seals on the Syrian coast.

2. The results of Gruvel's work 1929–1931

2.1 Pinnipeds

There is only one pinniped on the coasts of Syria and even then, it is quite rare—it is the monk seal (*Monachus albiventer*, Bodd.) =*Monachus monachus* (Hermann, 1779) which is found sporadically [4]. It has been seen, from time to time, some representatives frolicking more or less offshore. Since that date, monk seals have been in conflict with humans, because they feed on fish and compete with them in the catch. The fishermen hit it with spears to keep it away from the fish in the fishing nets. Since that date, monk seals have been in conflict with humans, because they feed on fish and compete with them in the catch. Thus, the monk seal has been subjected to disturbance and threat since that time, and other types of problems have been added to these threats that threaten the monk seal populations, such as urban expansion and investment of beaches, the increase in the density of tourists at sea, and the increase in marine transportation and fishing vessels. The natives, moreover, hunt these animals relentlessly when they see them, for they are extremely fond of their meats.

2.2 Cetaceans

All those who have traveled in the Mediterranean know the common dolphins *Delphinus delphis* (L., 1758) which ships, by making regular leaps out of the water, either while hunting, or simply, while playing. Among the other Delphinidae, he cites a number of forms that are only found sporadically on the Syrian coast, such as the "Blowers" (*Tursiops tursio*, 5 m long; the Grampus *Grampus griseus* Cuvier) fairly common throughout the Mediterranean and, finally, the *Globicephalus melas* (Flower) Greater Porpoise 7–8 m long. *Porpoises Phocoena communis* Cuv. are not uncommon either.

Among the Physeteridae, we can point out as making rare appearances in this part of the eastern Mediterranean [4], the largest of all—the Sperm whale (*Physeter macrocephalus* L.) which can reach 20–25 m in length and which is, in sum, fairly common in the Mediterranean. It is, moreover, one of the most cosmopolitan species and it is found everywhere, on all the seas. We know that it is through the large intestine of these large mammals that ambergris is secreted, under repeated excitation, produced on the intestinal walls, by the bites of the many beaks of Octopuses and other cephalopods, which these animals make their customary food.

A few years ago (before 1931), a Hispano-Norwegian cetacean hunting company was established near Algeciras (Spain). It hunted both sperm whales and, above all, whales, in the proportion of one of the former against 20 of the latter.

However, these large marine mammals are not confined to the approaches to the Strait of Gibraltar; they penetrate widely into the Mediterranean and have been observed fairly often in the eastern Mediterranean, off Port Said, and even beyond, on the coasts of Palestine and Syria, sporadically, however.

Among the other Physeteridae, he notes, in the Mediterranean, *Hyperoodon rostratus*, whose habitat is rather northern and its size does not exceed 10 m, another species *Zyphius cavirostris*, this is belonging to the Zyphidae family which is barely as long as the previous one and which like all the Physeteridae, feeds on octopuses and other cephalopods which it most often hunts in the deep sea. Homeric struggles occur between these large animals and octopuses of good size, from which the sperm whales emerge victoriously, it is true, but not without injuries, sometimes enormous, produced by the terrible hooks of the suction cups of the arms of certain large species abysmal.

The Baleinidae family is also represented, at least by two species. The largest in size is *Baloenoptera musculus* (Comp.) which the Norwegians designate under the name of "Finhval" which sometimes reaches up to 25 m in length and can weigh from 20 to 25,000 kg. Baleen can be nearly a meter long. This animal gives 60–70 hl of oil and 120–130 kg of baleen.

A much smaller species, often referred to as the dwarf whale, has also been reported in the Mediterranean—it is *Baloenoptera rostrata*, "waagehval" as called by the Norwegians, also referred to as the sharp-nosed whale, which normally lives in arctic regions, but often descends into the Bay of Biscay and, sometimes, into the Mediterranean.

These are the main species of Cetaceans that frequent the Mediterranean in general and which are found, sporadically, on the coasts of Syria. Their number is completely insufficient to justify the industrial exploitation of these mammals [4].

3. Results of stranding monitoring of marine mammals during 1991–2021

Data on stranding marine mammals have been collected in different areas along the Syrian coast (**Figure 2**). In each case, the site was documented, the stranding individual described and the taxonomic position determined using scientifically

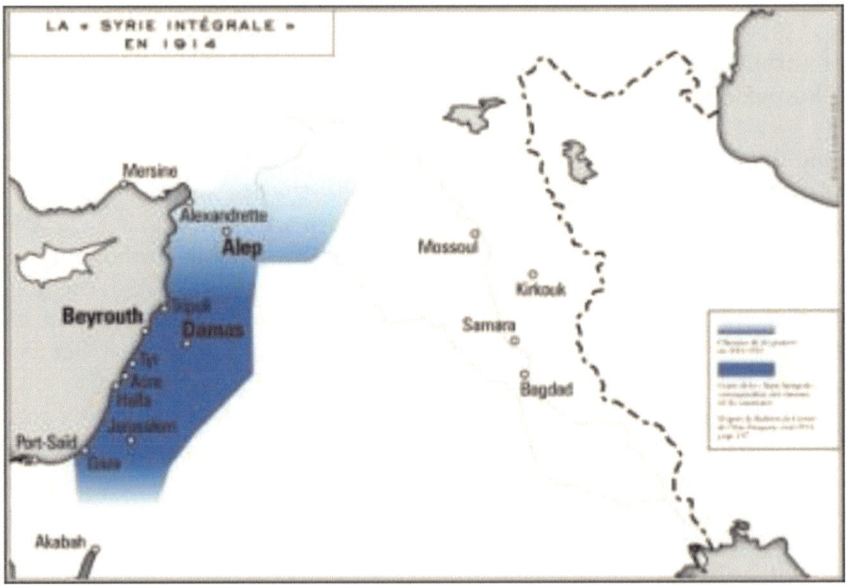

Figure 2.
The Integral Syrian coast where Gruvel made a survey of marine mammals during the period 1929–1931 (Source of the map: Andurain [5]).

Species	Sex	Date	Location	Fate	Source/note
Pseudorca crassidens		20/06/1991	South of Latakia	D	Skull found on the beach [8]
Stenella coeruleoalba	–	25/9/02	Albasit	D	Saad [10]
Tursiops truncates	–	02/10/02	North Latakia	Live	Filmed by Saad [10]
Tursiops truncates	F	03/03/03	Albasit	D	Saad [10]
Tursiops truncates		16/3/03	Latakia	D	Saad [10]
Ziphius cavirostris	M	11/3/2005	Om Altiour 50 km North of Latakia	D	495 cm, SSAEP
Megaptera novaeangliae	M	5/04/03	North Tartous	D	Male, 785 cm [9]
Tursiops truncates	–	25/11/2003	Tartous	D	Saad [10]
Delphinus delphi	–	17/03/2004	Banias	D	Saad [10]
Tursiops truncates	–	24/07/2004	Banias	D	Saad [10]
Physeter macrocephalus	M	25/4/2005	Tartous	D	1045 cm [10]
Delphinus delphie	–	16/9/2005	North Latakia	D	Saad [10]
Tursiops truncates	–	18/12/2005	Albasit	D	Saad [10]
Ziphius cavirostris	–	9/3/2006	Ibn Hani	D	Saad [10]
Tursiops truncates	M	8/10/2006	Ibn Hani	D	Floating adrift, male, tied with a rope, possibly after bycatch
Delphinus delphie	–	8/4/2007	Ibn Hani	D	A. Saad SSAEP
Tursiops truncates	–	12/4/2007	Harbor of Banias	D	Stranded dead, 315 cm, old age (based on worn-out teeth)
Tursiops truncates	–	4/05/2007	Ibn Hani 10 km N of Latakia	D	223 cm, likely killed by a large propeller (deep parallel cuts on the body)
Ziphius cavirostris	–	14/5/07	Jable	D	Stranding [10]
Grampus griseus	–	12/7/2007	South of Tartus	D	A long-decomposed corpse on land near the shore
Stenella coeruleoalba	F	8/9/07	Jable	Live	Recorded by A. Saad and SSAEP [10]
Stenella coeruleoalba	M	11/9/07	Jable	D	Stranding [10]
Stenella coeruleoalba	–	28/11/07	Albasit	D	Stranding [10]
Stenella coeruleoalba	–	16/12/07	Banias	D	Stranding [10]
Delphinus delphie	–	3/2/08	Albasit	D	Stranding, SAAEP
Stenella coeruleoalba	–	11/2/08	North Latakia	D	A. Saad SSAEP
Ziphius cavirostris	F	3/03/2008	South Jable (Rmielah)	D	SSAEP 290 cm (very young female)
Stenella coeruleoalba		15/4/08	Tartous	D	Saad [10]

Species	Sex	Date	Location	Fate	Source/note
Triosups Trincatus	–	18/10/2019	North of Latakia (Debjeat)	Live was returned to the sea *L* = 3 m	Informations provide by SSAEP
Balaenoptera acutorostrata	F	7/03/2020	North of Tartus (Alkhrab)	D	Str. L = 3 m, SSAEP

Table 1.
Cetacean records along the Syrian coast (1991–2021).

Species	Scientific name	Number of sighting	Mean group size	Max group size
Bottlenose dolphin	*Tursiops truncatus*	1	1	1
Common dolphin	*Delphinus delphis*	8	1	5
Risso's dolphin	*Grampeus griseus*	1	1	2
Striped dolphin	*Stenella coeruleoalba*	2	1	1
Sperm whale	*Physeter macrocephalus*	1	1	1
Unidentified whale		1	1	2

Table 2.
A summary of all cetacean sightings made during the ASI Okeanos in Syrian waters.

approved identification keys [6, 7]. This survey of cetaceans stranded along the Syrian coastline (2002–2021) and a review of the literature allowed us to record a total number of 30 stranding events from 1991 to 2021. They included three species of Balaenopteridae, one species of Physeteridae, one species of Ziphidae, five species of Delphinidae, and one species of Phocidae (**Tables 1** and **2**).

Recently, many marine mammal species have been observed along the Syrian coast may be for feeding or breeding, or migration behavior, but the most realistic reason, from our point of view, is to increase scientific monitoring along the Syrian coast, as well as to raise awareness of fishermen about the importance of reporting their sightings to the research team at the Syrian Society for the Protection of the Aquatic Environment and at Tishreen University, Which allowed the recorded of several stranding every year in different areas on the Syrian coast [8–12], as summarized them in **Table 1**, bearing in mind that there may be a few stranding that was not monitored, perhaps because of the remoteness of the area or the lack of reporting. Hence, our objective in this work is to compile and review the records and strandings of marine mammals along the Syrian coast and to provide further suggestions to protect these vulnerable species.

4. Results of a field survey onboard research ships Yunis S in August 2008

In partnership with a team from the Faculty of Fisheries—Istanbul University, within the framework of a joint Syrian–Turkish–Lebanese research program.

To determine the species composition, size, and distribution of the cetacean population, sighting data were collected during a research cruise carried out in summer (July 11–24, 2008) in the international water of the Eastern Mediterranean Sea, as well as the Turkish, Lebanese and Syrian territorial waters. This research cruise was conducted with a 32-m research vessel YUNUS—from the faculty of fisheries—at Istanbul University. The average speed of the vessel was 8–10 nautical miles/h. At each cetacean sighting, date and time, species, the group size of animals, location (coordinates), depth, sea state, and the behavior of animals were recorded by a single observer placed on the bridge deck.

Totally 860 nautical miles of survey effort were made and 16 sightings (108 animals) were recorded. During the study, 5 *Physeter macrocephalus* in one sighting, 72 *Stenella coeruleoalba* in five sightings, 2 *Grampus griseus* in one sighting (associated with *S. coeruleoalba* individuals), and 2 *Delphinus delphis* in one sighting, and 27 *Tursiops truncatus* in nine sightings were recorded. The overall encounter rate was 0.18 sightings/10 nm [13].

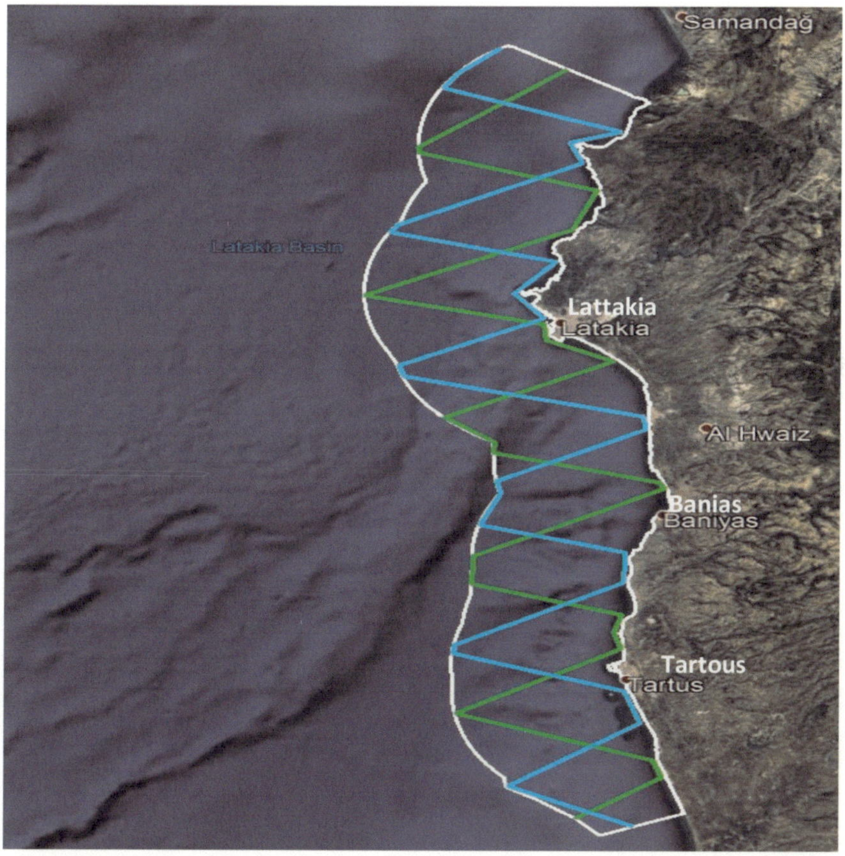

Figure 3.
Scheme of the field survey of marine mammals in the Syrian territorial waters (the tracks of the Okeanos boat along the Syrian coast) which was implemented within the ASI–ACCOBAMS project during the period from July 27 to August 8, 2019.

5. Results of a field survey on abundance and distribution of cetaceans in the Syrian waters of the ACCOBAMS Agreement

In the framework of participating in the work of the ASI[1] project launched at the Sixth Meeting of the Parties in ACCOBAMS[2] (Monaco, 22–25 November 2016), with the aim of creating an integrated, collaborative, and coordinated monitoring system of the state of cetaceans within the ACCOBAMS region (Black Sea, Mediterranean Sea, and Contiguous Atlantic Area), which was developed and implemented by the permanent secretariat of ACCOBAMS, in coordination and with the support of the riparian countries of the Mediterranean and local scientists, the author led a team of local experts. Researchers and trainees conducted visual surveys in Syrian territorial waters [14], where surveys were conducted onboard a ship chartered by the ASI Project called Okeanos, on two sets of perpendicular tracks on the coast (**Figure 3**) during the period: July 27 until August 8, and included 431 km of effort on track. The tracks cut a few miles from the Lebanese and Turkish borders. During the survey, five cetacean species were documented by sightings (**Table 2**) (**Figure 4**) [14, 15].

6. Monk seal

The Mediterranean monk seal (*Monachus monachus*) is a common mammal in the Syrian coastal water, especially in the northern parts of the Syrian coast near the Syrian–Turkish border. These animals live up to 45 years and their length is approximately 2.5 m. Male and female Mediterranean monk seals can be easily distinguished since the first ones have black color, while the second ones have brown fur. After

Figure 4.
*Pictures of five dolphins (*T. truncatus*) in front of the beach of Wadi Qandil, about 30 km north of Latakia, during the field survey process by direct viewing onboard the boat Okeanos within the framework of the ASI–ACCOBAMS project: 2 August 2019 [14].*

---

[1] Accobams Survey Initiative.

[2] Agreement on the Conservation of Cetaceans of the Black Sea, the Mediterranean Sea, and Contiguous Atlantic Area.

Figure 5.
A female Mediterranean monk seal was found carrying a full fetus that had mistakenly been killed (Port of Latakia, Syrian coast, 7/22/2013).

Figure 6.
An individual monk seal was seen swimming near the beach of Burj Islam, 20 km north of Latakia, 12 December 2021.

being considered "critically endangered" for the previous 19 years, the Mediterranean monk seal *Monachus monachus* (Hermann, 1779) has been classified as "Endangered," according to the IUCN Red List. Although the population is doing better in the Mediterranean, the monk seal remains a relatively rare species and difficult to observe with large consumption of fish per day, the animal has long been the bane of fishermen in the archipelago because they regularly tore their nets. Gradually, he succeeded in escaping their radars and by the same to those of the scientists who deplore the insufficiency of information. Hunted for a long time, we are now trying to get to know the animal better. Hope today lies in the development of new tools for the preservation of marine biodiversity.

These species are threatened by human activity and environmental pollution. More than 32 sightings of this animal have been recorded in the waters of the Syrian coast between 1996 and 2021. On July 22, 2013, a female Mediterranean monk seal was found carrying a full fetus (**Figure 5**), who had mistakenly killed someone who had infiltrated the main port with the aim of sabotaging. The victim's mother and her chick were placed in two fiberglass basins filled with formalin. On December 12, 2021,

Suborder	Family	Species	Scientific name	Reg/ vis	Gruvel 1931	2002–2020
Order Cetartiodactyla, Infraorder: Cetacea						
Odontocet	Delphinidae	Bottlenose dolphin	*Tursiops truncatus*	Regular	+	+
	=	Common dolphin	*Delphinus delphis*	Regular	+	+
	=	Risso's dolphin	*Grampeus griseus*	Regular	+	+
	=	Striped dolphin	*Stenella coeruleoalba*	Regular	+	+
	=	Long-finned pilot whale	*Globicephala melas*	Absent	+	–
	=	False killer whale	*Pseudorca crassidens*	Visitor	–	+
	Fhysteridae	Sperm whale	*Physeter macrocephalus*	Regular	+	+
Mysticeti	Balaenopteridae	Humpback whale	*Megaptera novaeangliae*	Visitor	+	+
	Balaenopteridae	Fin whale	*Balaenoptera physalus*	Visitor	+	+
	Balaenopteridae	Minke whale	*Balaenoptera acutorostrata*	Visitor	+	+
	Balaenopteridae	Blue whale	*Balaenoptera musculus*	Absent	+	–
	Balaenidae	North Atlantic right whale, is	*Eubalaena glacialis*	Absent	+	–
	Ziphidae	Cuvier's beaked whales	*Ziphius cavirostris*	Regular	+	+
Order carnivorous						
Pinnipedia	Phocidae	Mediterranean monk seal	*Monachus Monachus*	Regular	+	+

Table 3.
Comparison between the marine mammal species that were previously documented nearly a century ago and those whose existence has been documented in Syrian waters during the last two decades.

the presence of two seals was documented in a cave on the seashore in the Samra region on the Syrian–Turkish border, and one individual was seen swimming near the beach of Burj Islam, 20 km north of Latakia, where there are several caves believed that seals take refuge in them to rest or take care of their young (**Figure 6**). It is worth noting that the author recorded sounds of monk seals in a rocky cave south of Burj Islam on July 16, 2020, and these sounds indicate the presence of more than one individual. A seal swimming in the same place was previously documented in 2005 [16].

In addition, according to local residents and fishermen, there are eight seals that were seen intermittently roaming the waters between Burj Islam, Wadi Qandil (20 km north of Latakia), and Samra, on the Turkish–Syrian border.

Before the end of this chapter, we have compiled the results of previous and current work in **Table 3**, which shows a comparison between the species of marine that were previously documented nearly a century ago and those whose existence has been documented in Syrian waters during the last two decades.

7. Conclusion

In this work, marine mammals in the Syrian waters (the Levant Basin) and the changes that have occurred in their qualitative composition have been documented for about a hundred years now. Hence, it is of great importance, and the results showed that the Syrian waters and those of the neighboring countries have become a visiting area for many marine mammals that have not been previously observed, and more cases of stranding have been observed and documented, as this is attributed to the fragility of marine mammals in front of various human activities, such as increasing fishing operations associated with the diversity and efficiency of fishing gear, and the rise in commercial shipping and seismic activities. There are four species whose presence in the Syrian marine waters was mentioned a century ago, but neither alive nor dead have been seen in this area since 1996. The repeated sightings of many species (Lives or strandings) may be due to climatic changes which affect their migration and mobility from one place to another. Stakeholders should pay more attention to marine mammals in Syria through increased awareness, and the continuous monitoring, documentation, and mapping of recorded strandings to further suggest measures on how to protect such important and vulnerable species.

Acknowledgements

We thank the members of the Syrian Society for the Aquatic Protection who have made and are still exerting great efforts in monitoring stranded marine mammals on the Syrian coast, as well as the fishermen who cooperated with us by reporting many individuals of stranded whales and dolphins.

Conflict of interest

The authors declare no conflict of interest.

A. Annex: documentation of some stranding and sighting events

Humpback whale stranded in Tartous, March 12th, 2003.

Bottlenose dolphin in the sea 15 km North of Latakia, 1 October 2002.

Sperm whale stranded in Tartuous Beach, April 18th, 2005.

Bottlenose dolphin in Ibn Hani, 10 km North of Latakia, April 2006.

Striped dolphin in Shkaifat beach, 15 km South of Latakia, 9 September 2007. He was returned alive to the sea, and 2 days later he was seen dead on the beach 2 km north.

Bottlenose dolphin. At the port of catch landing in Latakia by-catch with nets, it was returned to the sea. 28 October 2019.

Risso's dolphin on the beach 12 km South of Tartus city July 2012.

Cuvier's beaked whales. In Ibn Hani Bay (10 km North of Latakia), 9 March 2006.

Cuvier's beaked whales in Alhamidea beach (20 km South of Tartus) July, 2010.

Minke whell. On Alkhrab beach (15 km North of Tartus city) 7 March 2020.

Author details

Adib Saad[1*] and Ilene Mahfoud[2*]

1 Marine Sciences Laboratory, Faculty of Agriculture, Tishreen University, Latakia, Syria

2 Syrian Society for Aquatic Environment Protection (SSAEP), Tishreen University, Latakia, Syria

*Address all correspondence to: adibsaad52@gmail.com; adibsaad@tishreen.edu.sy and ilenemahfoud@tishreen.edu.sy

References

[1] Giuseppe Notarbartolo di Sciara. Marine mammals in the Mediterranean Sea: An overview. Advances in Marine Biology. 2016;**75**:2-428. DOI: 10.1016/bs.amb.2016.08.005

[2] Lejeusne CP, Chevaldonné C, Pergent-Martini C, Boudouresque F, Pérez T. Climate change effects on a miniature ocean: The highly diverse, highly impacted Mediterranean Sea. Trends in Ecology & Evolution. 2010;**25**:250-260

[3] Bearzi, G, Fortuna CM. Common bottlenose dolphin *Tursiops truncatus* (Mediterranean subpopulation). In: The Status and Distribution of Cetaceans in the Black Sea and Mediterranean Sea (Ed. by R.R. Reeves & G. Notarbartolo di Sciara, compilers and editors), pp. 64–73, 2006. IUCN Centre for Mediterranean Cooperation, Málaga, Spain

[4] Gruvel A. LES ÉTATS DE SYRIE Richesses marines et fluviales Exploitation actuelle – Avenir. Maritimes et Coloniales: Société d'Éditions Géographiques; 1931

[5] La Grande-Syrie AJ. Diachronie d'une construction géographique (1912–1923). Revue des mondes musulmans et de la Méditerranée. 2017;**141**:33-35

[6] Shirihai H. Whales, Dolphins and Seals: A Field Guide to the Marine Mammals of the World. London: A and C Black Publishers Ltd. 2006. 384 pp. ISBN 978-0713670370

[7] Jefferson TA, Leatherwood S, Webber MA. FAO species identification guide. In: Marine Mammals of the World. Rome: FAO; 1993. 320 p. 587 figs

[8] Kasparek M. The false killer whale, *Pseudorca crassidens* (Cetacea: Delphinidae), new for Syria. Zoology in the Middle East. 1997;**14**:23-26

[9] Saad A, 2008. National overviews on the current status of cetacean-fisheries conflicts including bycatch and depredation with a critical review of any historical data in Syria (Eastern Mediterranean). Proceeding of International Workshop on Bycatch Within The ACCBAMS Area, FAO, Rome, 20 p

[10] Saad A. First record of a humpback whale stranding on the coast of Syria (Eastern Mediterranean). ACCOBAMS Fins Newsletter. 2004;**1**(1):10

[11] Gonzalvo J, Bearzi G. Action Plan for the Conservation of Cetaceans in Syria. Regional Activity Centre for Specially Protected Areas (RAC/SPA); 2008. 45 pp

[12] Saad A. 2017. Cetaceans species recorded on the Syrian coast during the period: 2002–2016 and perspective. 4th Biennial Conference—CSMC4, ACCOBAMS, Oran (Algeria), 11–13 Nov 2017

[13] Dede A, Saad A, Fakhri M, Öztürk B. Cetacean sightings in the Eastern Mediterranean Sea during the cruise in summer 2008. 2012

[14] Saad A, Ali A, Badran M, Ali M, Aidek A, Othman A, Dib S, Saleh M, Ahmad M, Badour F. 2019: The ACCOBAMS survey initiative project in Syrian marine water (ASI). ACCOBAMS, 2021. Estimate of abundance and distribution of cetaceans, marine mega-fauna and marine litter in the Mediterranean Sea from 2018–2019 surveys. In: Panigada S, Boisseau O, Canadas A, Lambert C, Laran S, McLanaghan R, Moscrop A, editors. ACCOBAMS–ACCOBAMS Survey Initiative Project. 27th July–8th August 2019; Monaco; 2019. 177 pp

[15] ACCOBAMS. Estimate of abundance and distribution of cetaceans, marine mega-fauna and marine litter in the Mediterranean Sea from 2018–2019 surveys. In: Panigada S, Boisseau O, Canadas A, Lambert C, Laran S, McLanaghan R, Moscrop A. Ed. ACCOBAMS–ACCOBAMS Survey Initiative Project, Monaco. 2021. 177 pp.

[16] Jony M, Ibrahim A. 2006. The first confirmed record for Mediterranean monk seals in Syria. Abstract, p. 54. In: UNEP/MAP, RAC/SPA. Report of the International Conference on Monk Seal Conservation. 17–19 September 2006; Antalya, Turkey. 69 p

Chapter 7

How Do Whales See?

Elena Vecino, Xandra Pereiro, Noelia Ruzafa and Sansar C. Sharma

Abstract

The eyes of two whales *Balaenoptera physalus* and *Baleoptera borealis* were studied by our group. In this chapter, we present the anatomical, histological, immunohistochemical and ultrastructural studies of the eyes of both types of whales. Based on the results, we can conclude that at least in these two species, the whales are rod monochromat; their resolution is very limited due to the reduced number of retinal ganglion cells, some of which were giant size (more than 100 micrometers in diameter). The excellent representation of melanopsinic positive retinal ganglion cells suggests an adaptation to the dim light as well as involvement in the circadian rhythms. The large cavernous body located in the back of the eye may provide a mechanism that allows them to move the eye forward and backwords; this may facilitate focusing and provide protection from cold deep-sea temperatures.

Keywords: whales, cetacean, eye, vision, retina, optic nerve, anatomy

1. Introduction

The cetaceans are fully aquatic mammals with 89 species recognized by the Society of Marine Mammalogy Committee on Taxonomy [1]. The artiodactylian ancestors moved to aquatic life about 50 million years ago [2] and at present some relatives, like hippopotamuses, are semi-aquatic.

The adaptation to the new media, under the water, induced several changes in the morphology of the eyes. Although anatomy and functional vision have been studied in the odontocetes like dolphins and orca (cetaceans with teeth), limited studies have been done on mysticete eyes, the larger cetaceans (with baleen). Moreover, very few studies have been done on the molecular distribution and ultrastructure of the retina. In the present chapter, we will summarize the main results of our group on the structure of the eyes of two large fin whales (18 meters large and 20 tons in weight) that beached on the Cantabric coast of Spain; they died a few hours after beaching, but this allowed us to study the eye in perfect morphological details and some biochemical analysis. Since the eyes of both fin whales had similar size and the animals were of the same sex, we will discuss both without distinction. Both specimens were adult males; one was *Balaenoptera physalus* (beached on the 4th of February 2019 in Sopelana beach, Biscay Spain) and the other *Baleoptera borealis* (beached on the 20th of January 2021 in Serantes, Asturias Spain).

The results shown in the present chapter are based on anatomical, histologic, immunohistochemical and electron microscopy studies performed during the last 2 years that have partially been published in several articles [3–5]. The methodology was explained in those articles. Here, we will concentrate on explaining the conclusions of how our research has helped us understand how the whales see.

2. Anatomy of the eye

The first anatomical discovery was the size of the eye. Even though it was very small in proportion to the large body size of the animal, the whale eye was huge. When compared with other terrestrial mammals, it weighs 1 kilogram and is almost 13 cm in diameter. The shape of the eyeball was markedly flattened on the anterior segment compared with the spherical shape of the terrestrial mammals. As a result, the axial length of the eyecup was smaller than its diameter is close to a hemisphere. The eyes were located, as in other mysticetes, in the oral commissure. Two very thick eyelids protect the eye. The eyelids lack eyelashes that would not function under water (**Figure 1A–D**). Due to the adaptation to the big body size, and to resist the high pressures, the sclera was very thick, and was 4 cm in the thickest posterior part of the eye. Its composition was mainly collagen indicating that even when its texture is like bone the composition lacks hydroxyapatite [3]. Encapsulated sensory corpuscles were found in the sclera in groups in the proximity of the iridocorneal angle. It has been suggested that this corpuscle within the cetacean eye may function as pressure receptors, possibly to control intraocular pressure [6, 7].

The cornea was an elongation of the sclera that becomes transparent only in the most anterior part of the eye. The thickness of the cornea was not uniform. The cornea was thinner in the centre and thicker at the periphery with a peripheral rim of almost 4 mm thick while the central part was 2 mm. Moreover, the cornea in the central part of the eye was flattened, another adaptation to vision underwater [3]. The cornea in the case of these cetaceans acted as a weak divergent lens and the refractive index is very little different from that of the water (**Figure 1C**). Therefore, in cetaceans, light refraction and focusing of an image on the retina are almost entirely performed by the lens [8]. The intraocular lens, the crystalline, in these mysticetes was almost spherical similar to that in fishes and measured 2 cm in diameter. It was placed in the centre of the eyecup so that light coming from any direction is focused on the retina. The curvature of the lens surface provides sufficient refractive power to focus images on the retina. Thus, the optics of these cetaceans are similar to those of fish, reflecting the adaptation to the common environment and optical properties of the vision under the water. The very viscous vitreous body filled the vitreous chamber and as in fish, this viscosity is an adaptation to equilibrate the refraction.

The adaptation of the cetacean vision, from the deep underwater where the light is very low to a rapid change of brightness at the surface of ocean when the animal dives into the well-lit water surface, is perhaps due to the iris structure and its pupil shape. Most cetaceans have a special pupil with a U-shaped slit, however, the two mysticetes studied had the pupil oval and horizontally elongated with dimensions of 5 x 3 cm (**Figure 1A**). Several studies on the refraction index in the cetacean eyes, and especially in dolphins, had evoked the conclusion that the refractive index between the air and the cornea should make the cetacean eye catastrophically myopic (near-sighted) in air. However, this myopia is countered by the presence of flattened cornea [9]. The constriction of the iris and pupil shape can provide a correction of aerial

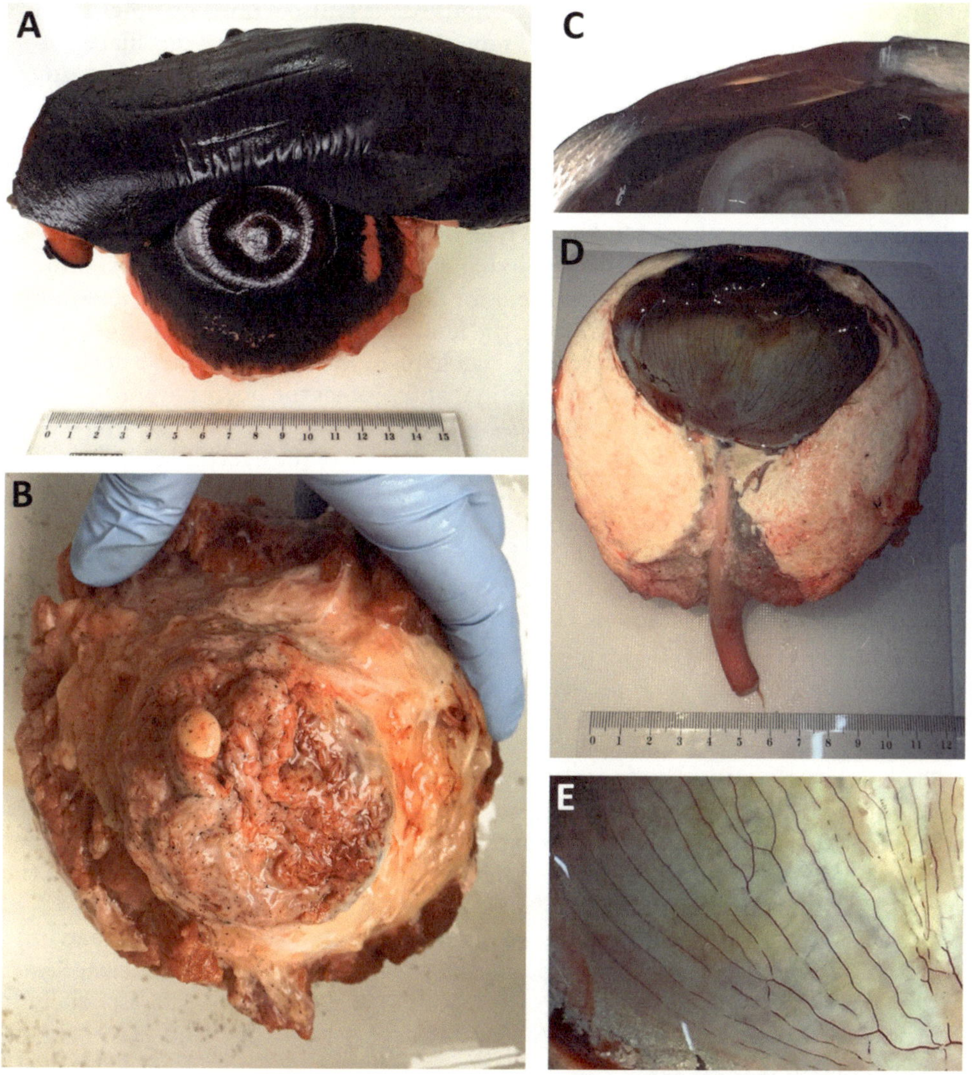

Figure 1.
(A) Eye and upper eyelid. Note the oval shape of the cornea and the round iris. (B) Detailed of the cavernous tissue surrounding the optic nerve. (C) Transversal section of the cornea and location of the round crystalline in the Centre. Note that the cornea is flat and thicker in the periphery than in the Centre. (D) Section of the eye showing the thick sclera, and the hemispheric retina and its vascularization. (E) Detail of the retina vascularization with parallel vessels.

myopia. Another adaptation of the cetacean eyes to conditions of low luminosity is a highly developed reflective layer, the tapetumlucidum, which lies behind the retinal pigment epithelium within the choroid. The tapetum in cetaceans has been described previously by [10]. We found a well-developed tapetum in the present two mysticetes studied. Its tapetum is formed with extracellular collagen fibrils and covers the complete fundus. What is unusual in the terrestrial mammals that live and hunt during the night is that the tapetumlucidum does not usually extend below the horizontal equator of the eye.

The eyes have strong extraocular muscles that cover a large cavernous tissue refilling the space that forms the sclera in the back of the eye and it surrounds the optic nerve with a conical shape. The cavernous tissue is also named by other authors as vascular plexus, rete ophthalmica and musculus retractor bulbi is speculated to function as a vascular rete to supply both heat and oxygen during dives [11]. A great number of blood vessels were surrounded by elastic fibers, smooth muscle and fat (**Figure 1B**). The function of this tissue could be to protect the optic nerve from the cold water and warm it and propel the eye outwards thereby helping with the focus. The massive musculus retractor bulbi, which produces axial displacements of the globe of the eye within the orbit, has been observed in other animals, mainly non-mammals like amphibians and the whale shark, and other cetaceans. However, the theory purposed by Kröger and Kirschfeld for dolphins in 1989, about the possibility that the focus could be based on shifting the lens backward due to the changes in the intraocular pressure, are very unprovable mechanism for whales, because of the very thick sclera (4 cm), that surrounds the retina, will protect the inner part of the eye from deforming and thus would be impossible to move the lens. We believe that the filling of the rete will increase necessary pressure to move the eyeball outwards. The retractor muscle will accompany the outward movement and when the rete is emptied the eye returns to the interior of the orbit. During this movement, the thick sclera will protect the eyeball from deforming forces. This could provide a mechanism to increase the field of vision when the eye protrudes. However, we do not believe that the lens will modify its position due to the propulsion of the eye.

3. The retina

The macroscopic view of the retina showed a dichotomous vascularization similar to that in a horse but in this case, it was holangiotic (**Figure 1D, E**). The vasculature had a radial distribution for the whole extension of the retina as previously described [12]. The shape of the retina is an incomplete hemisphere and a peripheral rim of the retina is bent inward (**Figure 1D**).

The thickness of the retina in cross-sections was not thicker than the human or pig retina. The number of rows of photoreceptors was 6–8 in the outer nuclear layer of the retina, however, the inner layers were thinner with less cellularity especially noticeable were the scarce and larger RGCs that were between 26.5 and 112.9 micrometers in diameter (**Figure 2A**) [4].

A detailed study of the photoreceptors showed that these species of mysticetes do not have cones. We used several antibodies to identify the different opsins, like rhodopsin, M/L opsin and S opsin as well as antibodies against rat and human cone arrestin. The only positive result was with rhodopsin, which specifically was present in rods. We conclude that in our study, the cone opsins that terrestrial animals used were not present in the whale retinas [4]. Previous studies demonstrated the monochromacy in some cetaceans and several mutations in the opsin gene sequences suggesting evolutionary modification of the cone cell function. In some cases, the cone structures have been partially maintained but with the absence of outer and inner segments [13]. Recent studies in other species of cetacean have found that cones have been adapted to other use, and instead of having visual pigments, like cone-opsins, they contain two proteins involved in magneto sensation, suggesting the possibility of an alternative functional role in responding to changes in geomagnetic fields [14].

Once the light impacts the photoreceptors, the next cells responsible for the transmission of the visual impulse within the retina are the bipolar and amacrine cells. We found that even when the cones' opsins were not present, bipolar and

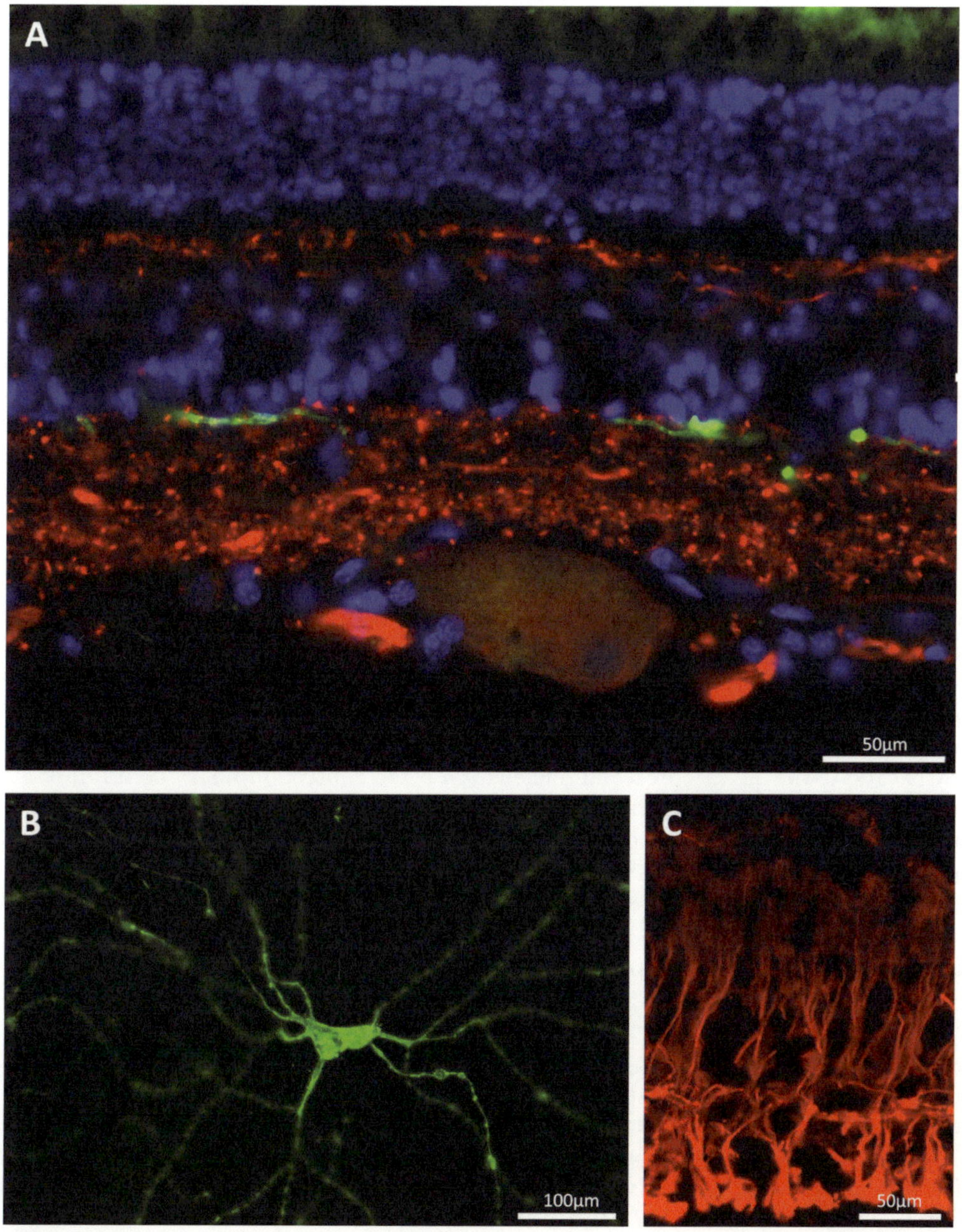

Figure 2.
(A) Immunostaining of a retina section. Red neurofilaments labels the large retinal ganglion cell and their branches. In blue DAPI, staining shows the 6–7 layers of photoreceptors in the upper part of the picture. (B) Flat mount of the retina showing a melanopsin positive retinal ganglion cell. (C) immunostaining of a retina section. In red, the Müller glial cells are stained with antibodies against vimentin.

amacrine cells maintained their molecular signals as well as location. The same results were found using electroretinograms to measure the electrical response to the light of these interneurons. [15] found that the cetaceans rod monochromat has both On and OFF bipolar cell pathways. The next step within the retina is the cells that are responsible to communicate the eye with the brain. These are the retinal ganglion cells (RGCs). The density of these cells was very low, lower than in any cetacean studied earlier [4], confirming the low resolution of that the whales' retina. It appears that relatively few RGCs may have to integrate the information from a great number of photoreceptors that are abundant. We did not count the number of photoreceptors, however, considering the large size of the eye and the number of rows of rods, the proportion of information that the RGCs have to integrate is massive. Even when the RGCs were scarce in number, the size of the cell bodies was 3 times larger than in humans, with some cells reaching 100 micrometers (**Figure 2A**).

Besides the RGCs that integrate the visual inputs and send this information to the visual centres of the brain for processing, other specific type of cells are located in the retinal ganglion cell layer that does not transmit information to the image forming visual area of the brain. These specific cells only transmit information regarding light intensity to the brain and are areas responsible for controlling the circadian rhythms and pupillary light reflexes. We have been the first research group to describe them in cetaceans. These cells are melanopsinipRGCs [4]. They form a mosaic that covers the complete area of the retina and were more abundant in the centre of the retina and reduce in density toward the periphery (**Figure 2B**). There are at least six types of ipRGCs and in the whale's retinas, we identify three types, the M1, M2 and M3 although the majority of them were M2. As per comparison in rodent retina, the number of M1 is higher than M2 or M3 [16]. In humans, M1d ipRGC is the predominant subtype [17]. These differences between species may be due to the different roles of the ipRGC subtypes. M1 projects to approximately 15 brain targets not involved in image forming activities, projecting predominantly to the suprachiasmatic nucleus (SCN) and to the olivary pretectal nucleus (OPN) to control pupillary light reflex [18, 19]. However, M2 ipRGCs that were the predominant ones in the whales′ retinas project mostly to the OPN and relatively fewer to the SCN suggesting that the control of the pupillary light reflex is very important for the whale [4].

In addition to neurons, there are glial cells in the retina that develop a very important role. The glia cells in cetacean's retinas were slightly different to other mammals. Thus, the astrocytes had a punctate pattern surrounding the large arteries of the retina but due to the very large and strong blood vessels, they were seldom visualized embracing the vessels as they typically do in other mammalian retinas. The Müller glia was very robust and with the same morphology as in other mammalian retinas (**Figure 2C**). They had the capacity to facilitate the elongation of the RGC axons [20] and for that reason, we had immortalized the whale's Müller glia for further study [5]. We also studied the microglial cells, which are the immune cells within the retina. They are present but were larger and more diffused. These differences with the terrestrial mammals could be related to the metabolism of the cetacean retina. For further discussion on the glial cells, see Ref. [4].

4. The optic nerve

The optic nerve was 1 cm in diameter, small in comparison with the large size of the eye. The axons in the optic nerve were mainly 'Giant' (greater than 15micrometres in diameter) corresponding to the very large RGCs found in the retina. The size of the axons and cells is a common plesiomorphic character of cetaceans rather than being related to the large body size of these animals. This interpretation is supported by the comparison with other large terrestrial species. Thus, in elephants there are no

Giant RGCs' despite the large size of the animals [21], indicating that this has been an adaptation to the aquatic environment.

The high astroglial content in the cetacean nerve could be due to the highly developed metabolic support that the central nervous system required to sustain nervous activity during anaerobic and energy-demanding tasks like prolonged apnea [22]. The bigger the diameter of the axons is, the faster the impulse travels [23]. There was considerably more astroglia in the cetacean optic nerve than inland mammals, with astrocyte processes and myelin occupying a higher proportion of the nerve [4].

5. Adaptation to the aquatic vision

Similar characteristics of adaptation to the aquatic vision as described in the present study, have been found in other mysticetes of smaller size humpback and Bryde's whales [24] or in Gray whale [25] aquatic mammals.

The described characteristics of the cornea play a minor role under the water because of its mall curvature and refractive indices of the media in front (seawater) and behind the cornea (aqueous humor). Although some refraction cannot be neglected especially due to the thickness [26]. The spatial resolving power of the cetaceans is very low compared to that of terrestrial mammals. Computing the peak of RGC density values range indicates that it is approximately 1 to 5 cycles/degree [24], compared for example with the horse that is approximately 20 cycles/degree [27], or the giraffe with approximately 20 cycles/degree [28].

Due to the large numbers of rod photoreceptors, and the few RGCs, it seems that for the cetaceans the sensitivity to the light is far more important than the resolution. A rod-vision provides better underwater vision in conditions where light intensity is low and light is scattered with increasing depth and rod dominance animals are more rapid to dark adaptation. Marine mammals use vision primarily in low light levels, where color vision may be of secondary importance. We believe that Peichl's comments are very appropriate in this regard 'the ocean for the whales is not blue' [29]. Moreover, the very well developed melanopsin RGCs network indicates that the sensibility to the changes in light is highly preserved and the control of the pupil is very important to reduce the amount of light that penetrates the retina thus, as a pinhole camera they could see better the object that it wants to see. The mechanism which protrudes the eye with the cavernous tissue together with the retractor bulbils muscle allows them to move the eye forward and backwords. This could facilitate the whales to focus when the eye is protruded outward and be protected from the cold temperatures when in the deep ocean waters where the eye is moved inside the orbit and covered by the eyelids.

6. Conclusions

The large whales' studies have been adapted to the underwater vision by a pupil that regulates the variations in luminosity together with the optic tapetum. The retina is similar to that of nocturnal terrestrial mammals with the absence of cones and large size of ganglion cells, separated by wide intercellular spaces, but a very well developed melanopsin system provides an adaptation for vision in low light environments. Thus, despite their eye large size, the retinal resolving power in cetaceans is generally weaker than in terrestrial mammals, due primarily to their low density of retinal ganglion cells.

Acknowledgements

This study was supported by ELKARTEK (KK-2019/00086), MINECO-Retos (PID2019-111139RB-I00), Grupos UPV/EHU (GIU 2018/150), and Proyectos de InvestigaciónBásica y/o Aplicada (PIBA_2020_1_0026) to EV, Basque Government postdoctoral grant (POS_2020_2_0031) to XP and UPV/EHU postdoctoral grant (ESPDOC20/058) to NR. We would like to thank the organizations AMBAR and CEPESMA for their support in getting the whales' eyes in the best possible conditions.

Conflict of interest

The authors declare no conflict of interest.

Author details

Elena Vecino[1]*, Xandra Pereiro[1], Noelia Ruzafa[1] and Sansar C. Sharma[2]

1 Experimental Ophthalmo-Biology Group (GOBE), Department of Cell Biology and Histology, University of Basque Country UPV/EHU, Leioa, Spain

2 Department of Ophthalmology, Cell Biology and Anatomy, New York Medical College, Valhalla, NY, USA

*Address all correspondence to: elena.vecino@ehu.es

References

[1] Committee on Taxonomy. List of marine mammal species and subspecies. [Internet]. 2021. Available from: www.marinemammalscience.org. [Accessed: 14-03-2022]

[2] Rose KD. Evolution. The ancestry of whales. Science. 2001;**293**(5538):2216-2217. DOI: 10.1126/science.1065305

[3] Vecino E, Ruzafa N, Pereiro X, Zulueta A, Sarmiento A, Díez A. The Extracellular Matrix of the Human and Whale Cornea and Sclera: Implications in Glaucoma and Other Pathologies. Extracellular Matrix-Developments and Therapeutics. Rijeka: IntechOpen; 2021

[4] Ruzafa N, Pereiro X, Vecino E. Immunohistochemical characterisation of the whale retina. Frontiers in Neuroanatomy. 2022;**16**:813369. DOI: 10.3389/fnana.2022.813369

[5] Pereiro X, Beriain S, Rodríguez L, Roiz-Valle D, Ruzafa N, Vecino E. Characteristics of whale Müller glia in primary and immortalized cultures. Frontiers in Neuroscience. 2022;**16**:1-16. DOI: 10.3389/fnins.2022.854278

[6] Miller S, Samuelson D, Dubielzig R. Anatomic features of the cetacean globe. Veterinary Ophthalmology. 2013;**16** (Suppl 1):52-63. DOI: 10.1111/vop.12025

[7] Vrabec F. Encapsulated sensory corpuscles in the sclerocorneal boundary tissues of the killer whale Orcinus orca L. Acta Anatomica (Basel). 1972;**81**(1):23-29. DOI: 10.1159/000143740

[8] Sivak JG. Accommodation in vertebrates: A contemporary survey. Current Topics in Eye Research. 1980;**3**:281-330

[9] Rivamonte A. Eye model to account for comparable aerial and underwater acuities of the bottlenosed dolphin. Netherlands Journal of Sea Research. 1976;**10**:491-498

[10] Dawson WW. The cetacean eye. In: Herman LM, editor. Cetacean Behavior: Mechanisms and Functions. New York: Wiley; 1980. pp. 53-100

[11] Ninomiya H, Yoshida E. Functional anatomy of the ocular circulatory system: Vascular corrosion casts of the cetacean eye. Veterinary Ophthalmology. 2007;**10**(4):231-238. DOI: 10.1111/j.1463-5224.2007.00544.x

[12] Pilleri G, Wandeler A. Ontogenesis and functional morphology of the eye of the finback whale, Balaenoptera Physalus Linnaeus (Cetacea, Mysticeti, Balaenopteridae). Acta Anatomica (Basel). 1964;**57**(Suppl. 50):1-74

[13] Schweikert LE, Fasick JI, Grace MS. Evolutionary loss of cone photoreception in balaenid whales reveals circuit stability in the mammalian retina. The Journal of Comparative Neurology. 2016;**524**(14):2873-2885. DOI: 10.1002/cne.23996

[14] Smith MA, Waugh DA, McBurney DL, George JC, Suydam RS, Thewissen JGM, et al. A comparative analysis of cone photoreceptor morphology in bowhead and beluga whales. The Journal of Comparative Neurology. 2021;**529**(9):2376-2390. DOI: 10.1002/cne.25101

[15] Collin SP, Davies WL, Hart NS, Hunt DM. The evolution of early vertebrate photoreceptors. Philosophical Transactions of the Royal Society of London. Series B, Biological Sciences. 2009;**364**(1531):2925-2940. DOI: 10.1098/rstb.2009.0099

[16] Lax P, Ortuno-Lizaran I, Maneu V, Vidal-Sanz M, Cuenca N. Photosensitive Melanopsin-containing retinal ganglion cells in health and disease: Implications for circadian rhythms. International Journal of Molecular Sciences. 2019;**20**(13):1-18. DOI: 10.3390/ijms20133164

[17] Ortuno-Lizaran I, Esquiva G, Beach TG, Serrano GE, Adler CH, Lax P, et al. Degeneration of human photosensitive retinal ganglion cells may explain sleep and circadian rhythms disorders in Parkinson's disease. Acta Neuropathologica Communications. 2018;**6**(1):90. DOI: 10.1186/s40478-018-0596-z

[18] Schmidt TM, Alam NM, Chen S, Kofuji P, Li W, Prusky GT, et al. A role for melanopsin in alpha retinal ganglion cells and contrast detection. Neuron. 2014;**82**(4):781-788. DOI: 10.1016/j.neuron.2014.03.022

[19] Schmidt TM, Chen SK, Hattar S. Intrinsically photosensitive retinal ganglion cells: Many subtypes, diverse functions. Trends in Neurosciences. 2011;**34**(11):572-580. DOI: 10.1016/j.tins.2011.07.001

[20] Garcia M, Forster V, Hicks D, Vecino E. Effects of muller glia on cell survival and neuritogenesis in adult porcine retina in vitro. Investigative Ophthalmology & Visual Science. 2002;**43**(12):3735-3743

[21] Stone J, Halasz P. Topography of the retina in the elephant Loxodontaafricana. Brain, Behavior and Evolution. 1989;**34**(2):84-95. DOI: 10.1159/000116494

[22] Mazzatenta A, Caleo M, Baldaccini NE, Maffei L. A comparative morphometric analysis of the optic nerve in two cetacean species, the striped dolphin (Stenellacoeruleoalba) and fin whale (Balaenopteraphysalus). Visual Neuroscience. 2001;**18**(2):319-325. DOI: 10.1017/s0952523801182155

[23] Lobsiger CS, Cleveland DW. Neurofilaments: Organization and function in neurons. In: Encyclopedia of Neuroscience. Vol. 6. United Kingdom: Elsevier; 2009. pp. 433-436

[24] Lisney TJ, Collin SP. Retinal topography in two species of baleen whale (Cetacea: Mysticeti). Brain, Behavior and Evolution. 2018;**92**(3-4): 97-116. DOI: 10.1159/000495285

[25] Mass AM, Supin AY. Ocular anatomy, retinal ganglion cell distribution, and visual resolution in the gray whale, Eschrichtiusgibbosus. Aquatic Mammals. 1997;**23**(1):17-28

[26] Kröger RHH, Kirschfeld K. Refractive index in the cornea of a harbor porpoise (Phocoenaphocoena) measured by two-wavelengths laser-interferometry. Aquatic Mammals. 1994;**20**:99-107

[27] Timney B, Keil K. Visual acuity in the horse. Vision Research. 1992;**32**(12):2289-2293. DOI: 10.1016/0042-6989(92)90092-w

[28] Coimbra JP, Hart NS, Collin SP, Manger PR. Scene from above: Retinal ganglion cell topography and spatial resolving power in the giraffe (Giraffacamelopardalis). The Journal of Comparative Neurology. 2013;**521**(9):2042-2057. DOI: 10.1002/cne.23271

[29] Peichl L, Behrmann G, Kroger RH. For whales and seals the ocean is not blue: A visual pigment loss in marine mammals. The European Journal of Neuroscience. 2001;**13**(8):1520-1528. DOI: 10.1046/j.0953-816x.2001.01533.x